SELECTED POEMS
1990

D. J. ENRIGHT
SELECTED POEMS
1990

Oxford   New York
OXFORD UNIVERSITY PRESS
1990

Oxford University Press, Walton Street, Oxford OX2 6DP

Oxford New York Toronto
Delhi Bombay Calcutta Madras Karachi
Petaling Jaya Singapore Hong Kong Tokyo
Nairobi Dar es Salaam Cape Town
Melbourne Auckland
and associated companies in
Berlin Ibadan

Oxford is a trade mark of Oxford University Press
© D. J. Enright 1990

This selection from Collected Poems 1987 first published as an
Oxford University Press paperback 1990

British Library Cataloguing in Publication Data
Enright, D. J. (Dennis Joseph), 1920–
Selected poems, 1990. (Oxford poets)
I. Title
821'.914
ISBN 0–19–282625–5

Library of Congress Cataloging in Publication Data
Enright, D. J. (Dennis Joseph), 1920–
[Poems. Selections]
Selected poems, 1990 / D.J. Enright.
p.   cm.
I. Title.
PR6009.N6A6   1990   821'.914—dc20   89-37189
ISBN 0–19–282625–5

Typeset by Wyvern Typesetting Ltd. Bristol
Printed in Great Britain by
J. W. Arrowsmith Ltd., Bristol

I am indebted to Shirley Chew of the University of Leeds for advice on the contents of this selection from my *Collected Poems 1987*.
  The sequences posed a slight problem. While there wasn't room to print them in their entirety, it seemed sad to omit them altogether. Consequently what is offered here is a fair representation in which the element of 'story', such as it is, has been preserved.

DJE

Contents

Waiting for the Bus

She hung away her years, her eyes grew young,
    And filled the dress that filled the shop;
Her figure softened into summer, though wind stung
    And rain would never stop.

    A dreaming not worn out with knowing,
A moment's absence from the watch, the weather.
    I threw the paper down, that carried no such story,
But roared for what it could not have, perpetual health
                    and liberty and glory.
    It whirled away, a lost bedraggled feather.

Then have we missed the bus? Or are we sure
    Which way the wind is blowing?

The Egyptian Cat

How harsh the change, since those plump halcyon days
                  beneath the chair of Nakht—
Poised in conscious honour above the prostrate fish,
No backstairs bones, no scraped and sorry skeleton,
                  no death's-head
From a toppled dustbin, but sacred unity of fish
                  and flesh and spirit—
With its stately fantail, its fiery markings like your own,
As tiger-god found good its tiger-offering . . .

For your seat is now among beggars, and neither man
                  nor cat is longer honoured,
As you with your lank ribs slink, and he
                  sprawls among his stumps.
Young boys, they say, lay their limbs on tram-lines
                  to enter this hard métier—
But you, maddened among the gross cars, mocked by
                  klaxons,
Lie at last in the gutter, merely and clearly dead . . .

And so I think of the old days—you, strong and
                  a little malignant,
Bent like a bow, like a rainbow proud in colour,
Tense on your tail's taut spring, at Thebes—
Where Death reigned over pharaohs, and by dark arts
Cast a light and lasting beauty over life itself—you,
Templed beneath the chair, tearing a fresh and virgin fish.

University Examinations in Egypt

The air is thick with nerves and smoke: pens tremble in
                              sweating hands:
Domestic police flit in and out, with smelling salts and aspirin:
And servants, grave-faced but dirty, pace the aisles,
With coffee, Players and Coca-Cola.

Was it like this in my day, at my place? Memory boggles
Between the aggressive fly and curious ant—but did I really
Pause in my painful flight to light a cigarette or swallow drugs?

The nervous eye, patrolling these hot unhappy victims,
Flinches at the symptoms of a year's hard teaching—
'Falstaff indulged in drinking and sexcess', and then,
'Doolittle was a dusty man' and 'Dr Jonson edited the Yellow
                              Book.'

Culture and aspirin: the urgent diploma, the straining brain—
                              all in the evening fall
To tric-trac in the café, to Hollywood in the picture-house:
Behind, like tourist posters, the glamour of laws and
                              committees,
Wars for freedom, cheap textbooks, national aspirations—

And, further still and very faint, the foreign ghost of happy
                              Shakespeare,
Keats who really loved things, Akhenaton who adored the
                              Sun,
And Goethe who never thought of Thought.

On the Death of a Child

The greatest griefs shall find themselves
                inside the smallest cage.
It's only then that we can hope to tame
                their rage,

The monsters we must live with. For
                it will not do
To hiss humanity because one human threw
Us out of heart and home. Or part

At odds with life because one baby failed
                to live.
Indeed, as little as its subject, is
                the wreath we give—

The big words fail to fit. Like giant boxes
Round small bodies. Taking up improper room,
Where so much withering is, and so much bloom.

The Laughing Hyena, after Hokusai

For him, it seems, everything was molten. Court-ladies flow
                              in gentle streams,
Or, gathering lotus, strain sideways from their curving boat,
A donkey prances, or a kite dances in the sky, or soars like
                              sacrificial smoke.
All is flux: waters fall and leap, and bridges leap and fall.
Even his Tortoise undulates, and his Spring Hat is lively as a
                              pool of fish.
All he ever saw was sea: a sea of marble splinters—
Long bright fingers claw across his pages, fjords and islands
                              and shattered trees—

And the Laughing Hyena, cavalier of evil, as volcanic as
                              the rest:
Elegant in a flowered gown, a face like a bomb-burst,
Featured with fangs and built about a rigid laugh,
Ever moving, like a pond's surface where a corpse has sunk.

Between the raised talons of the right hand rests an object—
At rest, like a pale island in a savage sea—
                              a child's head,
Immobile, authentic, torn and bloody—
The point of repose in the picture, the point of movement
                              in us.

Terrible enough, this demon. Yet it is present and perfect.
Firm as its horns, curling among its thick and handsome hair.
I find it an honest visitant, even consoling, after all
Those sententious phantoms, choked with rage and
                              uncertainty,
Who grimace from contemporary pages. It, at least,
Knows exactly why it laughs.

Life and Letters

I sat on the parapet, swinging my legs, close under
A luminous sky: a bright night city lay to my right:
Beneath me the seething trams, and a song, long and sad,
From a white café. And history—my own—oh nothing more
    portentous—
Pressed me both ways.

The near stars smelt of jasmine, and the moon—that huge
    fallafel—faintly of garlic.
Electric crickets sang. And bats displayed their talents
In rings around me, which I was too afraid
To fear. It was a time when superstitions drop away.
All day I walked under ladders, forgot to boil the milk.

For history—in the smallest sense—had fallen about me:
Held for a moment between those toppling towers,
Unable to understand, hatefully lost in cheerless ways, I sat
Suspended, dumbfounded, uneasily contained within my
    debris,
Bare above the hard road, the stiff steel, the tight-faced trams.
Two natives noticed me and jeered: a bored policeman
    sauntered up:
I went inside.

*

Which is why I try to write lucidly, that even I
Can understand it—and mildly, being loath to face the
    fashionable terrors,
Or venture among sinister symbols, under ruin's shadow.
Once having known, at an utter loss, that utter
    incomprehension
—Unseen, unsmelt, the bold bat, the cloud of jasmine,
Truly out of one's senses—it is unthinkable
To drink horror from ink, to sink into the darkness of words,
Words one has chosen oneself. Poems, at least,
Ought not to be phantoms.

The Monuments of Hiroshima

The roughly estimated ones, who do not sort well
    with our common phrases,
Who are by no means eating roots of dandelion,
    or pushing up the daisies.

The more or less anonymous, to whom no human idiom
    can apply,
Who neither passed away, or on,
    nor went before, nor vanished on a sigh.

Little of peace for them to rest in, less of them
    to rest in peace:
Dust to dust a swift transition, ashes to ash
    with awful ease.

Their only monument will be of others' casting—
A Tower of Peace, a Hall of Peace, a Bridge of Peace
    —who might have wished for something lasting,
Like a wooden box.

Tea Ceremony

The garden is not a garden, it is an
    expression of Zen;
The trees are not rooted in earth, then:
    they are rooted in Zen.
And this Tea has nothing to do with thirst:
It says the unsayable. And this bowl
    is no vessel: it is the First
And the Last, it is the Whole.

Beyond the bamboo fence are life-size people,
Rooted in precious little, without benefit of philosophy,
Who grow the rice, who deliver the goods, who
Sometimes bear the unbearable. They too
    drink tea, without much ceremony.

So pour the small beer, Sumichan. And girls,
    permit yourselves a hiccup, the thunder
Of humanity. The helpless alley is held by
    sleeping beggars under
Their stirring beards, and the raw fish curls
At the end of the day, and the hot streets cry
    for the careless scavengers.

We too have our precedents. Like those who
    invented this ceremony,
We drink to keep awake. What matter
If we find ourselves beyond the pale,
    the pale bamboo?

A Moment of Happiness

The river-bed is dry. And dry the flesh
Of the long-dead cat. Only the drained fur,
In feathers remaining, spreads a lonely fan.

The sober banners lie in stacks, like fallen leaves, and
Dry the tight lanterns in the phantom shop:
Empty of the light they wait: they thunder
   under the tread of the spider.
The hot iron of the railroad hisses in the air.

   It is early autumn,
Waiting between two festivals; the dumb sky thin and blue
   like egg-shell.
Till the frugal women, the little aunts bent double
With dry aches, shall suck the heated saké
From their brittle claws.

This bareness pleases me, this hard dry air.
The feast will come. The candle flare
In the paper skull; the skull will grow rosy
And warm; plump fingers ring the cups.
And meanwhile no one talks of national rebirth,
   and no one talks of literary renaissances.

Blue Umbrellas

'The thing that makes a blue umbrella with its tail—
How do you call it?' you ask. Poorly and pale
Comes my answer. For all I can call it is peacock.

Now that you go to school, you will learn how we call
          all sorts of things;
How we mar great works by our mean recital.
You will learn, for instance, that Head Monster
          is not the gentleman's accepted title;
The blue-tailed eccentrics will be merely peacocks;
          the dead bird will no longer doze
Off till tomorrow's lark, for the letter has killed him.
The dictionary is opening, the gay umbrellas close.

          Oh our mistaken teachers!—
It was not a proper respect for words that we need,
But a decent regard for things, those older creatures
          and more real.
Later you may even resort to writing verse
To prove the dishonesty of names and their black greed—
To confess your ignorance, to expiate your crime,
          seeking a spell to lift a curse.
Or you may, more commodiously, spy on your children,
          busy discoverers,
Without the dubious benefit of rhyme.

The Individual

One of nature's errors, a grasshopper
Livid and long, greener than any grass.
       Yet his vivid song
Shakes the breathless morning, trusting
That men will think him a useless flower,
The animals only notice a precious stone.

Hokusai's Mad Poet

He dances on his naked native toe,
And stars and blots and jottings sport about his head,
Read or unread, his works lie in a silken pile.
Beneath the unpacific sky he dances, while
The autumn pine drops leaves of thought about his head.

One agile line creates him in his twisted robe,
From toe to glorious grin and balding top—
Apply for special status, in return for quip and quirk?
Straightened shall be your crooked line, and stopped your
                    hop!
Pick up your careless leaves and sort your thoughts,
Replace the stars in heaven, modify that smirk!

                    It is alleged
An empty saké bottle in your company was seen.
No more than drunk you are, on Old Japan—
So far as we can tell, undemocratic Poet San.

Kyoto in Autumn

Precarious hour. Moment of charity and the
　　　　less usual love.

Mild evening. Even taxis now fall mild.
Grey heart, grey city, grey and dusty dove.

Retiring day peers back through paper windows;
　　　　here and there a child
Digs long-lost treasure from between her feet.

Where yesterday the sun's staff beat,
　　　　where winter's claws tomorrow sink,
The silent rag-man picks his comfort now.

The straitened road holds early drunkards
　　　　like a stronger vow;
The season's tang renews the burning tongue.

Poetic weather, nowhere goes unsung,
However short the song. A pipe's smoke prints
Its verses on the hand-made paper of that sky;

And under lanterns leaping like struck flints,
　　　　a potter's novice squats
And finds his colours in the turning air.

A pallid grace invests the gliding cars.
The Kamo keeps its decent way, not opulent nor bare.
The last light waves a fading hand. Now fiercer seasons
　　　　start like neon in the little bars.

Changing the Subject

I had suggested, in exasperation, that he find
    something other to write about
Than the moon, and flowers, and birds, and temples,
    and the bare hills of the once holy city—

People, I proposed, who bravely push their way
    through the leprous lakes of mud.
It was the wet season, rain upon spittle and urine,
    and I had been bravely pushing my way.

It happened my hard words chimed with a new slogan,
    a good idea, since ruined—
'Humanism'. So I helped on a fashion, another like
    mambo, French chanson, and learning Russian.

Now he comes back, my poet, in a different guise:
    the singer of those who sleep in the subway.
'Welcome you are,' his vagrants declaim to each other,
    'a comrade of the common fate.'

'Are they miners from Kyushu?' he asks, these 'hobos
    all in rags.' And adds that
'Broken bamboo baskets, their constant companions, watch
    loyally over their sleeping masters.'

Thus my friend. He asks me if he has passed the test,
    is he truly humanistic,
Will I write another article, about his change of heart?
    I try to think of the subway sleepers.

Who are indescribable. Have no wives or daughters to sell.
    Not the grain of faith that makes a beggar.
Have no words. No thing to express. No 'comrade'.
    Nothing so gratifying as a 'common fate'.

Their broken bamboo baskets are loyal because no one
    would wish to seduce them.
Their ochre skin still burns in its black nest, though a
    hundred changed poets decide to sing them.

'Are they miners from Kyushu?' Neither he nor I will
  ever dare to ask them.
For we know they are not really human, are as apt themes
  for verse as the moon and the bare hills.

The Noodle-Vendor's Flute

In a real city, from a real house,
At midnight by the ticking clocks,
In winter by the crackling roads:
Hearing the noodle-vendor's flute,
Two single fragile falling notes . . .
But what can this small sing-song say,
Under the noise of war?
The flute itself a counterfeit
(Siberian wind can freeze the lips),
Merely a rubber bulb and metal horn
(Hard to ride a cycle, watch for manholes
And late drunks, and play a flute together).
Just squeeze between gloved fingers,
And the note of mild hope sounds:
Release, the indrawn sigh of mild despair . . .
A poignant signal, like the cooee
Of some diffident soul locked out,
Less than appropriate to cooling macaroni.
Two wooden boxes slung across the wheel,
A rider in his middle age, trundling
This gross contraption on a dismal road,
Red eyes and nose and breathless rubber horn.
Yet still the pathos of that double tune
Defies its provenance, and can warm
The bitter night.
Sleepless, we turn and sleep.
Or sickness dwindles to some local limb.
Bought love for one long moment gives itself.
Or there a witch assures a frightened child
She bears no personal grudge.
And I, like other listeners,
See my stupid sadness as a common thing.
And being common,
Therefore something rare indeed.
The puffing vendor, surer than a trumpet,
Tells us we are not alone.

Each night that same frail midnight tune
Squeezed from a bogus flute,
Under the noise of war, after war's noise,
It mourns the fallen, every night,
It celebrates survival—
In real cities, real houses, real time.

Displaced Person Looks at a Cage-Bird

Every single day, going to where I stay
    (how long?), I pass the canary
In the window. Big bird, all pranked out,
Looming and booming in the window's blank.

Closing a pawky eye, tapping its hairy chest,
    flexing a brawny wing.
Every single day, coming from where I stay
(How long?), I pass this beastly thing.

How I wish it were dead!
    —Florid, complacent, rent-free and over-fed,
Feather-bedded, pensioned, free from wear and tear,
Earth has not anything to show less fair.

I do wish it were dead!
    Then I'd write a better poetry,
On that poor wee bird, its feet in the air,
An innocent victim of something. Just like me.

The Poor Wake Up Quickly

Surprised at night,
The trishaw driver
Slithers from the carriage,
Hurls himself upon the saddle.

With what violence he pedals
Slapbang into the swarming night,
Neon skidding off his cheekbones!
Madly he makes away
In the wrong direction.
I tap his shoulder nervously.
Madly he turns about
Between the taxis and the trams,
Makes away electric-eyed
In another wrong direction.

How do I star in that opium dream?
A hulking red-faced ruffian
Who beats him on his bony back,
Cursing in the tongue of demons.
But when we're there
He grumbles mildly over his wage,
Like a sober man,
A man who has had no recent visions.
The poor wake up quickly.

Am Steinplatz

Benches round a square of grass.
You enter by the stone that asks,
                'Remember those whom Hitlerism killed'.
'Remember those whom Stalinism killed',
Requests the stone by which you leave.

This day, as every other day,
I shuffle through the little park,
                from stone to stone,
From conscience-cancelling stone to stone,
Peering at the fading ribbons on the faded wreaths.

At least the benches bear their load,
Of people reading papers, eating ices,
Watching aeroplanes and flowers,
Sleeping, smoking, counting, cuddling.
                Everything but heed those stony words.
They have forgotten. As they must.
Remember those who live. Yes, they are right.
                They must.

A dog jumps on the bench beside me.
Nice doggie: never killed a single Jew, or Gentile.
Then it jumps on me. Its paws are muddy, muzzle wet.
Gently I push it off. It likes this game of war.

At last a neat stout lady on a nearby bench
Calls tenderly, 'Komm, Liebchen, komm!
Der Herr'—this public-park-frau barks—
                'does not like dogs!'

Shocked papers rustle to the ground;
Ices drip away forgotten; sleepers wake;
The lovers mobilize their distant eyes.
                The air strikes cold.
There's no room for a third stone here.
                I leave.

Saying No

After so many (in so many places) words,
It came to this one, No.
Epochs of parakeets, of peacocks, of paradisiac birds—
Then one bald owl croaked, No.

And now (in this one place, one time) to celebrate,
One sound will serve.
After the love-laced talk of art, philosophy and fate—
Just, No.

Some virtue here, in this speech-stupefied inane,
To keep it short.
However cumbrous, puffed and stretched the pain—
To say no more than, No.

Virtue (or only decency) it would have been,
But—no.
I dress that death's-head, all too plain, too clean,
With lots of pretty lengths of saying,

No.

The Quagga

By mid-century there were two quaggas left,
And one of the two was male.
The cares of office weighed heavily on him.
When you are the only male of a species,
It is not easy to lead a normal sort of life.

The goats nibbled and belched in casual content;
They charged and skidded up and down their concrete
   mountain.
One might cut his throat on broken glass,
Another stray too near the tigers.
But they were zealous husbands; and the enclosure was
   always full,
Its rank air throbbing with ingenuous voices.

The quagga, however, was a man of destiny.
His wife, whom he had met rather late in her life,
Preferred to sleep, or complain of the food and the weather.
For their little garden was less than paradisiac,
With its artificial sun that either scorched or left you cold,
And savants with cameras eternally hanging around,
To perpetuate the only male quagga in the world.

Perhaps that was why he failed to do it himself.
It is all very well for goats and monkeys—
But the last male of a species is subject to peculiar pressures.
If ancient Satan had come slithering in, perhaps . . .
But instead the savants, with cameras and notebooks,
Writing sad stories of the decadence of quaggas.

And then one sultry afternoon he started raising Cain.
This angry young quagga kicked the bars and broke a camera;
He even tried to bite his astonished keeper.
He protested loud and clear against this and that,
Till the other animals became quite embarrassed
For he seemed to be calling them names.

Then he noticed his wife, awake with the noise,
And a curious feeling quivered round his belly.
He was Adam: there was Eve.
Galloping over to her, his head flung back,
He stumbled, and broke a leg, and had to be shot.

Monkey

Once again the Year of the Monkey is here.
I was born in the Year of the Monkey—
Surely a fellow can talk about himself a bit,
    in his own year?

Monkeys are like poets—more than human.
Which is why they do not take us very seriously.
Not to be taken seriously is rather painful.
    To a corner of my cage I retired, mysteriously,
And had sad thoughts. (They may have been deep.)
Big eyes damp with a semi-permanent tear, my thin hands
    held my heavy head from tumbling into sleep.

But even boredom bored me. I hurled myself from bar
To bar, swung on my aching tail, gibbered and grunted
    with all the expected lack of finesse.
If you do a thing at all, do it well. So I hunted
    avidly for the fleas which in fact I do not possess.

Exercise is always good. Daruma lost his legs
    through meditating overlong and deleteriously:
Now they call him a saint, and use him as a paperweight.
'Thank God for Monkey,' thus spake a kindly spectator,
    backing towards the gate,
'Monkey saves us from taking ourselves too seriously!'

Whereupon he straightened his tie, murmured something
    about an important committee . . .
You can have too much exercise. I went back to my niche,
    put on my thinking face, sad and full of pity.
Proudly refused a banana. All night suffered hunger's itch.

But in the next town—who can tell?
    You may like to know that in Chinese
My name (though I cannot write it, being less well-informed
   than some of the flock)
Bears the meaning, 'Monkey comes to Town'.

I think you will find my name on our posters—down
    there, just below the performing pekinese.
Somebody ought to tell the town I'm coming.
    Next year belongs to the Cock.

In Memoriam

How clever they are, the Japanese, how clever!
The great department store, Takashimaya, on the
Ginza, near Maruzen Bookshop and British Council—
A sky-scraper swaying with every earth-tremor,
Bowing and scraping, but never falling (how clever!).
On the roof-garden of tall Takashimaya lives an
Elephant. How did he get there, that clever Japanese
Elephant? By lift? By helicopter? (How clever,
Either way.) And this young man who went there to teach
(Uncertificated, but they took him) in Tokyo,
This Englishman with a fine beard and a large and
(It seemed) a healthy body.
                                        And he married an orphan,
A Japanese orphan (illegitimate child of
A geisha—Japanese for 'a clever person'—and a
Number of customers), who spoke no English and
He spoke no Japanese. (But how clever they were!)
For a year they were married. She said, half in Japanese,
Half in English, wholly in truth: 'This is the first time
I have known happiness.' (The Japanese are a
Clever people, clever but sad.) 'They call it a
Lottery,' he wrote to me, 'I have made a lucky dip.'
(She was a Japanese orphan, brought up in a convent.)
At the end of that year he started to die.
They flew him to New York, for 2-million volt treatment
('Once a day,' he wrote, 'Enough to make you sick!')
And a number of operations. 'They say there's a
90% chance of a cure,' he wrote, 'Reversing
The odds, I suspect.' Flying back to his orphan,
He was removed from the plane at Honolulu and
Spent four days in a slummy hotel with no money or
Clothes. His passport was not in order. (Dying men
Are not always clever enough in thinking ahead.)
They operated again in Tokyo and again,
He was half a man, then no man, but the cancer
Throve on it. 'All I can say is,' he wrote in November,
'Takashimaya will damned well have to find
Another Father Christmas this year, that's all.'
(It was. He died a week later. I was still puzzling
How to reply.)

                    He would have died anywhere.
And he lived his last year in Japan, loved by a
Japanese orphan, teaching her the rudiments of
Happiness, and (without certificate) teaching
Japanese students. In the dungeons of learning, the
Concentration campuses, crammed with ragged uniforms
And consumptive faces, in a land where the literacy
Rate is over 100%, and the magazines
Read each other in the crowded subways. And
He was there (clever of them!), he was there teaching.
Then she went back to her convent, the Japanese
Widow, having known a year's happiness with a
Large blue-eyed red-bearded foreign devil, that's all.
There is a lot of cleverness in the world, but still
Not enough.

Reflections on Foreign Literature

The stories which my friends compose are very sad.
They border on the morbid (which, in the literatures
Of foreign languages, we may licitly enjoy, for they cannot
                  really
Corrupt, any more than we can be expected to discriminate).

(Sometimes I ask myself: Do I live in foreign countries
Because they cannot corrupt me, because I cannot be
Expected to make the unending effort of discrimination?
The exotic: a rest from meaning.)

('The officer shall engage in no activities whatsoever
Of a political nature,' says my contract, 'in the area where he
                serves.'
And all activity, it seems, is political.)

Anyway, the stories of my friends are very sad.
I am afraid they are largely true, too, discounting the
                grace-notes of my elegant friends.
At the heart of the ideogram is a suffering man or woman.

I remember my friend's friend, a barmaid in Shinjuku, at a
                literary pub—
Neither snowy-skinned nor sloe-eyed (though far from
                slow-witted),
Neither forward nor backward, of whom my friend
(A former PEN delegate) said in a whisper:
'Her life-story would make a book. I shall tell you one day . . .'
The day never came. But I can imagine the story.

My friend's friend also made special ties out of leather;
My friend gave me one, as a parting gift, a special memory of
                his country.
It has an elegant look; but when I wear it, it chafes my skin;
Whispering that nothing is exotic, if you understand, if you stick
                your neck out for an hour or two;
That only the very worst literature is foreign;
That practically no life at all is.

Apocalypse

'After the New Apocalypse, very few members were still in possession
of their instruments. Hardly a musician could call a decent suit his own.
Yet, by the early summer of 1945, strains of sweet music floated on the
air again. While the town still reeked of smoke, charred buildings and
the stench of corpses, the Philharmonic Orchestra bestowed the
everlasting and imperishable joy which music never fails to give.'
  (from *The Muses on the Banks of the Spree*, a Berlin tourist brochure)

It soothes the savage doubts.
One Bach outweighs ten Belsens. If 200,000 people
Were remaindered at Hiroshima, the sales of So-and-So's
New novel reached a higher figure in as short a time.
So, imperishable paintings reappeared:
Texts were reprinted:
Public buildings reconstructed:
Human beings reproduced.

After the Newer Apocalypse, very few members
Were still in possession of their instruments
(Very few were still in possession of their members),
And their suits were chiefly indecent.
Yet, while the town still reeked of smoke etc.,
The Philharmonic Trio bestowed, etc.

A civilization vindicated,
A race with three legs still to stand on!
True, the violin was shortly silenced by leukaemia,
And the pianoforte crumbled softly into dust.
But the flute was left. And one is enough.
All, in a sense, goes on. All is in order.

And the ten-tongued mammoth larks,
The forty-foot crickets and the elephantine frogs
Decided that the little chap was harmless,
At least he made no noise, on the banks of whatever river
                              it used to be.

One day, a reed-warbler stepped on him by accident.
However, all, in a sense, goes on. Still the everlasting and
imperishable joy
Which music never fails to give is being given.

Confessions of an English Opium Smoker

In some sobriety
I offer to recall those images:
Damsel, dome and dulcimer,
Portentous pageants, alien altars,
Foul unimaginable imagined monster,
Façades of fanfares, Lord's Prayer
Tattooed backwards on a Manchu fingernail,
Enigma, or a dread too well aware,
Swirling curtains, almond eyes or smell?

And I regain these images:
Rocked by the modern traffic of the town,
A grubby, badly lighted, stuffy shack—
A hollow in some nobody's family tree,
The undistinguished womb of anybody's
Average mother. And then me,
In all sobriety, slight pain in neck and back,
Expecting that and then a little more,
Right down to bed-rock.
This was no coloured twopenny,
Just a common people's penny sheet—

To read with cool avidity.
(What would you do with dulcimers,
And damsels, and such embarrassments?
Imagined beasts more foul than real monsters?
No man at peace makes poetry.)
Thus I recall, despite myself, the images
That merely were. I offer my sedate respects
To those so sober entertainments,
Suited to our day and ages.

The Burning of the Pipes

Bangkok, 1 July 1959

Who would imagine they were government property?—
Wooden cylinders with collars of silver, coming
From China, brown and shiny with sweat and age.
Inside them were banks of dreams, shiny with
Newness, though doubtless of time-honoured stock.
They were easy to draw on: you pursed your lips
As if to suckle and sucked your breath as if to
Sigh: two skills which most of us have mastered.

The dreams themselves weren't government property.
Rather, the religion of the people. While the state
Took its tithes and the compliance of sleepers.
Now a strong government dispenses with compliance,
A government with rich friends has no need of tithes.

What acrid jinn was it that entered their flesh?
For some, a magic saucer, over green enamelled
Parks and lofty flat-faced city offices, to
Some new Tamerlane in his ticker-tape triumph—
Romantics! They had been reading books.
Others found the one dream left them: dreamless sleep.

As for us, perhaps we had eaten too much to dream,
To need to dream, I mean, or have to sleep.
For us, a moment of thinking our thoughts were viable,
And hope not a hopeless pipe-dream; for us
The gift of forgiveness for the hole in the road,
The dog we ran over on our way to bed.
Wasn't that something? The Chinese invented so much.

A surprise to find they were government property
—Sweat-brown bamboo with dull silver inlay—
As they blaze in thousands on a government bonfire,
In the government park, by government order!
The rice crop is expected to show an increase,
More volunteers for the army, and navy, and
Government service, and a decrease in petty crime.

Not the first time that fire destroys a dream.
Coca-Cola sellers slither through the crowd; bats
Agitate among the rain-trees; flash-bulbs pop.
A holocaust of wooden legs—a miracle constated!
Rubbing his hands, the Marshal steps back from
The smoke, lost in a dream of strong government.
Sad, but they couldn't be beaten into TV sets;
As tourist souvenirs no self-respecting state
Could sponsor them, even at thirty dollars each.

Parliament of Cats

The cats caught a Yellow-vented Bulbul.
Snatched from them, for three days it uttered
Its gentle gospel, enthroned above their heads.
Became loved and respected of all the cats.
Then succumbed to internal injuries.
The cats regretted it all profoundly,
They would never forget the wrong they had done.

Later the cats caught a Daurian Starling.
And ate it. For a Daurian Starling is not
A Yellow-vented Bulbul. (Genuflection.)
Its colouring is altogether different.
It walks in a different, quite unnatural fashion.
The case is not the same at all as that of
The Yellow-vented Bulbul. (Genuflection.)

The kittens caught a Yellow-vented Bulbul.
And ate it. What difference, they ask, between
A Yellow-vented Bulbul and that known criminal
The Daurian Starling? Both move through the air
In a quite unnatural fashion. This is not
The Yellow-vented Bulbul of our parents' day,
Who was a Saint of course! (Genuflection.)

Dreaming in the Shanghai Restaurant

I would like to be that elderly Chinese gentleman.
He wears a gold watch with a gold bracelet,
But a shirt without sleeves or tie.
He has good luck moles on his face, but is not
    disfigured with fortune.
His wife resembles him, but is still a handsome woman,
She has never bound her feet or her belly.
Some of the party are his children, it seems,
And some his grandchildren;
No generation appears to intimidate another.
He is interested in people, without wanting to
    convert them or pervert them.
He eats with gusto, but not with lust;
And he drinks, but is not drunk.
He is content with his age, which has always suited him.
When he discusses a dish with the pretty waitress,
It is the dish he discusses, not the waitress.
The table-cloth is not so clean as to show indifference,
Not so dirty as to signify a lack of manners.
He proposes to pay the bill but knows he will not be
    allowed to.
He walks to the door like a man who doesn't fret
    about being respected, since he is;
A daughter or granddaughter opens the door for him,
And he thanks her.
It has been a satisfying evening. Tomorrow
Will be a satisfying morning. In between
    he will sleep satisfactorily.
I guess that for him it is peace in his time.
It would be agreeable to be this Chinese gentleman.

The Abyss

He walks towards the abyss.
Because, he's told, he owes it to himself.
And because he owes it something,
Because he is drawn by depth and darkness.

So many people walking this road!
Most will turn off a little further on.
But a message comes: To the office please,
Letters await your signature.

He likes signing letters. So returns.
But the abyss still calls. Louder
Than the voice of an air-conditioner.
So he retraces that boring stretch of road.

But is called back again: to compose
A reference for a student, one who left
Well before his time. The files record
'An average student', and nothing more.

This is a challenge to be met,
A challenge to the imagination's power.
The abyss will have to wait. Though he hears
Its voice, like a waterfall heard by a child.

So he retraces that boring stretch of road.
Something inside him starts to lick its lips.
He cannot frame an unpainted picture,
But he frames a rough blurb and a fine critique.

And here at last is the end of the road!
—It is the end of the month as well,
There are cheques to be written out.
Can't they wait? He admits they could.

But a crowd presses at the abyss.
Dangling their plumb-lines, fishing
For samples of its sides and bottom.
No use hanging about. He'll come back later.

News

It was all recounted, quite some time ago.
'There was my wife, clothed in a hundred patches',
'Corpses are piled on the grass, the smell is terrible',
'Blue is the smoke of war, white the bones of man'.

An old story, left to beginners to cut their teeth on,
To a conventional aside in a colour piece,
Or to Charitable Appeals. What we read about
Is another sort of war, more intellectual, more
Our sort of thing. The real war underneath the war.

Over a squabbling little country—the papers tell us—
A plane belonging to some foreign power shoots at
(And misses) the aircraft of another foreign power.
This constitutes an international incident.

The corpses, well and truly hit, on the grass,
The women in patches, the white hungry children—
This is not an international incident; hardly
A national one. Just an old unreadable story.
We are surprised Tu Fu should have dug it up.

In the Catalogue

It was a foreign horror.
A cold and lonely hour,
A place waste and littered,
And this figure standing there.

Like at first a prized
Cherry sapling swathed in straw.
It was no tree. It was enclosed
In a straw cocoon, and

Wore a hood of sacking
Over the might-be head
And the should-be shoulders.
It seemed to be looking.

What did I fear the most?
To ignore and bustle past?
To acknowledge and perhaps
Find out what best was lost?

It didn't accost. I did.
Rattling in my outstretched hand,
I hoped that money would talk,
A language of the land.

Some inner motion stirred the straw.
My stomach turned, I waited
For its—what?—its rustling claw
Or something I could not conceive.

What happened was the worst.
Nothing. Or simply, the straw
Subsided. 'Please, please!'
I begged. But nothing more.

Fear is glad to turn to anger.
I threw the money down and left,
Heedless of any danger,
Aside from vomiting.

From twenty yards I turned
To look. The shape stood still.
Another ten yards, and I strained
My eyes on icy shadows—

The shape was scrabbling for my coins!
I thanked my stomach. Then
Thanked God, who'd left the thing
Enough to make a man.

To Old Cavafy, from a New Country

'Imperfect? Does anything human escape
That sentence? And after all, we get along.'

But now we have fallen on evil times,
Ours is the age of goody-goodiness.

They are planning to kill the old Adam,
Perhaps at this moment the blade is entering.

And when the old Adam has ceased to live,
What part of us but suffers a death?

The body still walks and talks,
The mind performs its mental movements.

There is no lack of younger generation
To meet the nation's needs. Skills shall abound.

They inherit all we have to offer.
Only the dead Adam is not transmissive.

They will spread their narrowness into space,
The yellow moon their whitewashed suburbs.

He died in our generation, the old Adam.
Are our children ours, who did not know him?

We go to a nearby country, for juke-boxes and
Irony. The natives mutter, 'Dirty old tourists!'

We return, and our children wrinkle their noses.
Were we as they wish, few of them would be here!

Too good for us, the evil times we have fallen on.
Our old age shall be spent in disgrace and museums.

Prime Minister

Slowly he ticks off their names
On the long list:
All the young political men.

As he was once himself.
He thinks of how he despised the others
    —the apolitical,
       the English-educated,
       the students he called 'white ants
In their ivory tower'.

Not so long ago, in fact,
He coined that happy phrase 'white ants'.
How he despised them, all they cared for
Was lectures, essays and a good degree!

A small thing these days
    —he tells himself—
To be arrested.
Incredulously he remembers
Not once was he arrested, somehow.

Slowly he ticks off the names
On the list to be arrested.
Tonight, isn't it? Yes,
Between 2 and 4 when the blood runs slow.
The young political men,
Full of fire, hot-blooded.
    —For a moment,
He thinks he sees his own name there.
'Red ants,' he hisses,
Thrusting the list at a waiting policeman.

Misgiving at Dusk

In the damp unfocused dusk
Mosquitoes are gathering.

Out of a loudspeaker
Comes loud political speaking.
If I could catch the words
I could not tell the party.
If I could tell the party
I would not know the policy.
If I knew the policy
I could not see the meaning.
If I saw the meaning
I would not guess the outcome.

It is all a vituperative humming.
Night falls abruptly hereabouts.
Shaking with lust, the mosquitoes
Stiffen themselves with bloody possets.
I have become their stews.
Mist-encrusted, flowers of jasmine glimmer
On the grass, stars dismissed from office.

A Liberal Lost

Seeing a lizard
Seize in his jaws
A haphazard moth,

With butcher's stance
Bashing its brainpan
Against the wall,

It was ever your rule
To race to the scene,
Usefully or not.

(More often losing
The lizard his meal, not
Saving the moth.)

Now no longer.
Turning away, you say:
'It is the creature's nature,

He needs his rations.'
And in addition
The sight reminds you

Of that dragon
Watching you with jaws open
(Granted, it is his nature,

He needs his rations),
And—the thing that nettles you—
Jeering at your liberal notions.

Elegy in a Country Suburb

To strike that special tone,
Wholly truthful, intimate
And utterly unsparing,

A man communing with himself,
It seems you need to be alone,
Outwardly unhearing—

As might be now,
The streets wiped clean of traffic
By the curfew

(Apart from odd patrolling
Jeeps, which scurry through
This decent district),

The noise of killing
Far away, too distant
To be heard, above this silence

(A young Malay out strolling
—If you insist on instance—
Chopped by a Chinese gang of boys;

A party of Malays
Lopping an old man's Chinese head,
Hot in their need to burn his hut;

The riot squad,
Of some outlandish race,
Guns growing from their shoulders),

Until tomorrow's news,
And subsequent White Papers,
Analysing, blaming, praising,

Too distant to be heard
Above this heavy hush
Of pealing birds and crickets wheezing,

Tones of insects self-communing,
Birds being truthful with themselves,
Intimate and unsparing—

But birds are always chirping,
Insects rattling, always truthful,
Having little call to twist,

Who never entertained large dreams
Or made capacious claims,
Black lists or white lists:

It hardly even seems
The time for self-communing,
Better attend to nature's artists.

Small Hotel

Not *Guest*—
The Chinese, those corrected souls, all know
A guest is never billed, whereas the
Essence of my aspect is, I pay—

But *Occupier*. Good words cost no more.
The Occupier is hereby kindly warned,
It is forbidden strictly by the Law
—In smudged ungainly letters on his door—
Not to introduce into this room
Prostitutes and gambling, and instruments of
Opium Smoking and spitting on the floor.

By Order, all the lot, *The Management*.
The Chinese have immense respect for Order,
They manage anything you name except

To keep their voices down. Outside my door
The Management all night obeys the Law,
Gambles and introduces prostitutes,
And spits upon the floor and kicks around
The instruments of opium smoking.

It is forbidden to the Occupier
To sleep or introduce into his room
Dreams, or the instruments of restoration.
He finds he has his work cut strictly out
To meet the mandates of the Law and Order.

Coffee, frying garlic and a sudden calm
Imply the onset of a working day.
Kings and queens and jacks have all departed,
Mosquitoes nurse their bloody hangovers.

So large a bill of fare, so small the bill!
A yawning boy bears off my lightweight bag,
Sins of omission make my heavier load.
Insulting gringo. Cultural-imperialist.

Maybe a liberal tip will mollify
The Law, the Order, and the Management?—
With what I leave behind on that hard bed:
Years off my life, a century of rage
And envy.

Visiting

Mixing briefly with some
Who've lived for months, years, ages,
Deep down in the abyss,
On lower ledges,

He finds them easy,
Understanding, even almost gay.
How can this be?
He feels the ground slipping away
From under his feet.

So likeable, so considerate,
Yes, even almost healthy
(All their suicides unsuccessful).
How on earth can this be?

Unless they're visitors from above,
From the gayer, easier flatland?
And he the old dweller in abysses
—Apprehensive, prudent, pained—
On one of the upper ledges?

Well, that would explain
The odd resentment they arouse
In him. And the faint ancient pain
Of drawing breath. And his ragged nails.

Poet wondering what he is up to

—A sort of extra hunger,
Less easy to assuage than some
—Or else an extra ear

Listening for a telephone,
Which might or might not ring
In a distant room

—Or else a fear of ghosts
And fear lest ghosts might not appear,
Double superstition, double fear

—To miss and miss and miss,
And then to have, and still to know
That you must miss and miss anew

—It almost sounds like love,
Love in an early stage,
The thing you're talking of

—(but Beauty—no,
Problems of Leisure—no,
Maturity—hardly so)

—And this? Just metaphors
Describing metaphors describing—what?
The eccentric circle of your years.

What became of What-was-his-name?

Roughly once a quarter
I think of M—.
He must have been inside
For about three years now.
And three months longer
Each time I think of him.

Not that I knew him well.
But he used to hang around the place,
Looking for odd scraps of news.
Not a bad newspaperman,
I used to think,
A graduate with some ability
To express himself,
Comparatively well-mannered,
And properly sceptical
About the official hand-outs.
He would ask you what *you* thought.

Others have gone in
And come out
Since then,
Have taken new jobs and
Fathered new babies.
L—and F—are still inside,
But they're hard-core,
Prison is their favourite freedom.
M—wasn't that sort,
He seemed to me a food and drink man.
But he must have done something very bad.
The papers said nothing about it.

One could ask.
But it would draw attention to you,
And to M—.
They say the cases are reviewed
Every twelve months,
And it wouldn't do him any good
For the Special Branch to submit,
'He was asked after by . . .'

Like the police
In that other country:
'What would a gentleman like you
Want with those fellows?'
No use to answer:
'Well, it's only that
I've been teaching them English
For the past two or three years,
And . . .'
No one believes in pure friendship,
Charity went out
When Aid came in.

'Remember, this isn't Oxford or
Cambridge, you know.'
And then they might ask M—:
'Why would a foreigner be
Interested in you?
What favours have you done this
Foreigner?'
M—had better be totally uninteresting.

Is M—married, I wonder?
I can't remember whether I ever knew.
His wife would have news of him,
But she'll have returned to her family,
To her village.
It's not done
To pursue people to their villages.

Funny,
After three years
A new generation hangs around the place,
Hardly one of them has heard of M—.
It makes you feel your age.

But one thing's certain,
M—is alive.
They don't shoot people hereabouts,
They need to review them every year.
M—is a sort of pet, I suppose,
I remember him every three months.
Not that I ever knew him well.

Processional

for William Walsh

Where are they all?—
The Chancellor and the Vice-Chancellor,
The Deputy Vice-Chancellor and the Registrar,
The Bursar, the Deans of the several Faculties,
The Director of Extra-Mural Studies,
The Estate Officer and the Librarian,
The Chairman of the University Council,
The Esquire Bedell and the Public Orator?

For the scaffolding has collapsed, the
Scaffolding of the impending Science Tower
Has collapsed, with the long thin noise of the
Crumbling of a termite-riddled ivory tower.
And underneath are two female labourers,
Sought for by their colleagues like buried and
Perishable treasure. Now a trousered leg is
Uncovered, and pulled upon, and at its end is an
Ivory visage, a whitened stage concubine's,
Slashed with a vulgar wash of red.

And where is the University Health Physician?
(He is sick, he has left, he is on sick leave.)

And first will arrive the Fire Brigade,
With their hoses and helmets and hatchets
To exhume the already exhumed. And then
The police car with its mild unworried policemen
And hypnotic radio. And last of all,
In accordance with protocol, the ambulance,
To remove whatever the firemen and the policemen
Are no longer interested in.

But where are they all?—
The Development and the Public Relations Officers
And the various Assistant Registrars,
The Vice-Dean and the Sub-Dean of Law,
The Chairman of the Senior Common Room Committee,
The Acting Head of the School of Education,

The Readers and the Senior Lecturers
(The Professors we know are all at work),
And the Presidents respectively of the Local and
The Expatriate Academic Staff Associations?
This is a bad day for an accident.

Till a clerk arrives, a clerk from the
Administration, to administer the matter.
And order is imposed and sense is made.
The scaffolding consisted of old wood left out
Too long in the monsoon rains, and the women
Took too much sand up with them, because the
Contractor told them to get a move on, since he
Was hurrying to finish the job, because . . .
And they fell through four floors,
Carrying the scaffolding with them at each floor,
The sodden planking and the bamboo poles,
And now the scaffolding and the sand and the
Labourers all lie scattered on the ground.
The day and the hour are determined, and the
Victims identified as One Science Tower
(Uncompleted) and Two Labourers, Female, Chinese,
Aged about 20 and 40 respectively,
Who also look rather incomplete.

The Chancellor and the Vice-Chancellor,
The Deputy Vice-Chancellor and the Registrar . . .
There was little occasion for them after all.
The accident has been thoroughly administered—
Moved and seconded, carried and minuted.
A gaggle of idle Assistant Lecturers tap
On their watches, seditiously timing
The ambulance. And in the distance
The fire engine's bell can be heard already.

Brief Briefing

Colourful? It's green here,
Green and green. And then
A spectrum of ideologies—

Where too much colour
(Discernible red, let's say)
Spells detention.

And too much green
Means jungle here again.
They cut back both.

The spectrum fades to grey,
And grey apartment houses
Elbow out the jungle.

But then the flags of washing
Streaming out from poles—
It's all the colours here!

The Mysterious Incident
at the Admiral's Party

Moored in his favourite Eastern port,
This jolly British Admiral
Must give a party on his ship,
With jolly guests, Malays, Chinese,
And Indians and English too.
Says he, 'I like the sarong well,
Trim gear, I wear it when I can.'
Good Jack, who likes Malay costume!
Approval flutters like the gulls,
Down go the drinks, up come the words.
'Although,' he says, 'it tends to slip,
It slips and slides below my hips,
It's hard to keep a sarong up.'
A Chinese lady speaks, sedate
And sweet. 'Then Admiral you need
A songkok.' Songkok as you know
Is headwear proper to sarong.
But Admiral and nearby guests,
They do not know what songkok is,
They think they hear some other words.
Some gape, some giggle and some gasp,
And jolly shaken Jack withdraws
Upon his bridge, and all disperse.
This Chinese lady at a loss,
She asks her spouse in Mandarin,
But what, but why? Who, unbeknown,
Now scouts about the huddled groups.
Then joins his wife. 'Ah me, my dear,'
He murmurs in their tongue, 'To keep
His sarong up—they thought you said—
The Admiral needs a strong—'
'For shame!' in spotless Mandarin
This well-bred lady cries, 'Oh filthy-
Minded foreign hounds! Oh deep disgrace!
What can they think of Chinese dames,
These British gentlemen? Away!'

So Admiral is hurt, Malays
Offended, English persons shocked,
And Chinese lady hates the lot.
Weigh anchor, jolly Admiral—
Let drop these oriental tricks,
Be stayed with buttons and gold braid!

Cultural Freedom

Set free
From all committee,
What would you write?

One writes despite,
In spite of failure,
Of failing light.

One works because
Of lack of 'eisure;
Out of loss

Of liberty;
To fill deficiency
With presentness.

You need defeat's sour
Fuel for poetry.
Its motive power
Is powerlessness.

Back

Where is that sought-for place
Which grants a brief release
From locked impossibilities?
Impossible to say,
No signposts point the way.

Its very terrain vague
(What mountainside? What lake?)
It gives the senses nothing,
Nothing to carry back,
No souvenir, no photograph.

Towards its borders no train shrieks
(What meadowland? What creeks?)
And no plane howls towards its heart.
It is yourself you hear
(What parks? What gentle deer?).

Only desperation finds it,
Too desperate to blaze a trail.
It only lives by knowing lack.
The single sign that you were there is,
You know that you are back.

Children Killed in War

A still day here,
Trees standing like a lantern show,
Cicadas, those sparse eaters, at their song,
The eye of silence, lost in soundlessness.

And then, no warning given,
Or if foreseen, then not to be escaped,
A well-aimed wind explodes,
And limbs of trees, which cannot run away,
May only hide behind each other.

Grant their death came promptly there,
Who died too soon,
That pain of parting was not long,
Roots ready to let fall their leaves.

The wind burns out,
The trees, what's left, resume their stand,
The singers stilled, an iron comb
Wrenched roughly through their lives.

While you, your thinking blown off course,
Design some simple windless heaven
Of special treats and toys,
Like picnic snapshots,
Like a magic-lantern show.

Goodbye Empire

It had to go
So many wounded feelings
And some killings—
In a nutshell, too expensive

Though its going
Scarcely set its subjects free
For freedom—
Life still exacts a fee

In wounded feelings
And also killings
Slates wiped clean
Soon attract new chalkings

At least the old regime
Allowed its odd anomalies
Like my orphaned Irish dad
One of those Wild Geese

Who floundered over India
In the shit and out of it
Getting a stripe, and then
Falling off his horse and losing it.

Board of Selection

No, it is not easy to effect an
Appointment in English literature.
The chairman of the board is worrying over
Last year's riots (the bodies were traced
By the smell . . .) and next year's budget
(The British are pulling out . . . but leaving
Their literature behind them).

The dean of the faculty is an
Economist of repute and utility.
But what is the difference, he asks,
Between prose and poetry?
The candidate proposes adroitly that
Poetry is more economical than prose,
Viz., it says as much in half the space.

The economist is not satisfied.
In half the space, he muses . . . but
It takes him four times as long to understand
A piece of poetry as a piece of prose—
Which means . . .

The board make hasty calculations . . .
Which means that poetry is a false economy,
More haste, less speed.
The chairman remembers he has to build a nation
By the end of the month.

Neither I nor the candidate dare ask the
Esteemed economist which particular piece of
Poetry has so discomfited him.
It would probably have to do with daffodils
And this is an orchid-exporting country.
We submit that quite a lot of literature is
Prose and prose is pertinent to the economy.

A business man wants to know when the
Middle Ages stopped and the Renaissance started.
No one is sure. The director of education
Asks why there is so much sex in modern literature.
Because, the candidate ventures, there is so much
In modern life (excluding the English Department).
The chairman is brooding over the birth-rate.

Finally, after a disgruntled pilgrimage to
Canterbury and a brief stopover in 1984,
It is recommended that the candidate be offered
A temporary assistant lectureship at the bottom
Of the scale, subject to the survival of the
Economy, the nation, the university, the department,
And her hopes of completing her master's degree.

More Memories of Underdevelopment

> 'God's most deep decree
> Bitter would have me taste: my taste was me.'

A lapsed Wesleyan, one who dropped out
Halfway through the Wolf Cubs, and later ran howling
From Lourdes by the first bus back, whose idea of
High wit is 'God si Love', who would promptly
Ascribe the sight of Proteus rising from the sea
To spray in the eyes or alcohol in the brain—
Yet these words appal me with recognition,
They grow continuously in terror.

So how much more must they mean
To these young though ageless Catholics, to whom I am
Rashly expounding Father Hopkins!

But no,
They seem to find it a pleasing proposition,
The girls are thinking how sweet they taste, like moon-cake
Or crystallized pears from Peking,
The boys are thinking how good they taste, like crispy
      noodles
Or bird's-nest soup.

The poor old teacher is muttering curses,
Four letters cut to three out of care for his job:
Was I born to bring a sense of sin among you,
You oriental papists! Obviously Rome was not built in
Three hundred years. A lurching humanist,
Is it for me to instruct you in the fall complete?

Their prudent noses wrinkle almost imperceptibly.
Oh yes, they tell themselves, the poor old man,
His taste is certainly him . . .
And they turn to their nicer thoughts,
Of salted mangoes, pickled plums, and bamboo shoots,
And scarlet chillies, and rice as white as snow.

Master Kung at the Keyboard

for Lee Kum Sing

He's Oriental, he's Japanese, he's Chinese
Watch and you'll see him trip over his tail!
He's a child! What can he know of Vienna woods
Of Ludwig's deafness and J.S.B.'s fine ears?

Of tiaras and galas and programmes
Of hussars and cossacks and pogroms
Of Vespers, Valhallas and Wagrams
And the fine old flower of the Vienna woods?

(Wine, beef, pheasant, cheese, thirst, hunger)

Reared on rice and Taoist riddles
Water torture and the Yellow River
Yang Kwei-fei and one-stringed fiddles—
What can he know of the Water Music
Of barges and gondolas
Of emperors and haemophilias
Of the Abbé, the Princess, and her black cigars?

Wer das Dichten will verstehen
Muss ins Land der Dichtung gehen
Seven days with loaded Canon
Snapping prince and priest and peon.

So he went overseas for his studies?—
It is not in his blood.

What is in his blood?
Blood is.

(Rice, tea, pork, fish, hunger, thirst)

Compared with the minimum of 4,000 characters
Required at the finger-tips for near-literacy
And admission into provincial society
88 keys are child's play.

Play, child!

His heart pumps red rivers through his fingers
His hands chop Bechsteins into splinters
His breath ravishes the leaves
His hair never gets in his eyes.

I am down on my knees.

Every second pianist born is a Chinese
Schubert, Chopin, Mozart, Strauss and Liszt—
He'll be playing on
When the old Vienna woods have gone to chopsticks
Chopsticks every one.

Valediction

Into your roomy ear I speak,
Your lofty eye will read my lips—

Here's comfort:
Seeing I could never fill you
How should my going leave you empty?

My local lizard
Will hunt across the desk no longer,
Peer from under papers, and
Into the typewriter leap in close pursuit
Of enemy ants—
The desk will not be there.
But he'll be hunting still, a man
Of many dimensions, to whom you offer
Rich terrain.

Peevish bats will squirm in the eaves
To the south; to the north the starlings
Still hold assembly.

The shrew with her questing nose
Shall wheel across the tiles,
Less cover for her then, and thus
The less circuitous her quest. She
Never quested after me.

The swarming bees will piss
Their pollen as before;
Honey was merely meant to be,
Not meant for me.

For cats and dogs
Less of comfort, surely,
But more of interest.
Mosquitoes will miss me most,
The lizards will miss the mosquitoes.

The chain of being is thickly woven,
It does not break if one small link
Drops out. Nature makes do.

The *kempeitai* once came—
And used you as a torture chamber
Or, some oldtimers say, an officers' brothel—
The *kempeitai* then went.
I mention only one
Of your more notable tenancies.

It is not my going
You need to fear, old house,
But the coming of the bulldozer.
The new men have no respect.

Memoirs of a Book Reviewer

A month much like any other
There were five weekends in it
And no printing strike, but a rape
In Vietnam (I believe it was), a
Symbolical German dentist removing teeth
And illusions, and a Jewish Italian
Afflicted with accidie in an elegant style
(Or maybe that was the month before)
W. Pater wrote letters to an oriental maiden
Raped by GIs, while a bored Italian Jew
Discovered the source of eternal energy in
The Antarctic (Prix des Libraires) which was
Discovered a week later by a Swede in a
Documentary novel and a balloon. A number of
Quaint peasants murdered their husbands or
Wives etc. and got off scot-free because they
Were life-enhancing. And I forgot my name
(It was found at the bottom of a page)
I wrote elegant letters to an asian woman
Murdered by the foreign soldiery. Shakespeare
Turned up, with an apt word for everything
Especially titles. I forgot my title
(The postman brought me proofs of existence)
Much drinking of *vins de pays* went on. I can
Remember a hangover. And lots of sex
In a lost world at the bottom of the Antarctic
Discovered by E. Pound, an American balloonist
And I lost my memory for several days, but
Found it at the bottom of a cheque. In a
Spare hour I tried to write a poem on life and
Death in a Vietnamese hamlet called SW18
Ravaged by peace and the brutal schoolchildren
W. Pater came about the cooker
At some stage an exhaustive life of E. Pound
Appeared. A peasant named Confucius was raped
By the brutal intelligentsia. A large man came
About the rates, a book came about the rapes, a
Large book came about Shakespeare from Voltaire

To Ungaretti. An index slipped, trouble with
An appendix, my teeth hurt. At one point (I am
Almost sure) a book reviewer died at his desk
Whose name will be found at the bottom

In the Basilica of the Annunciation

A custodian, or guard, or butler
Reproves a youngish couple, their arms
Looped loosely, chastely round each other:
'This is no place for love.'

I ask myself
What this place is a place for.
This place is a building built by all the nations
This place must accommodate
(Without offence to any)
All the nations.

Its glories are the glories of the
Ideal Home Exhibition; the ark
Before the animals spilt drool and
Dung on its fragrant planks; a
Palace completed on the eve of revolution.
Many mansions do not make a house.

Is it a fact that in the streets
Camels are skipping through the eyes of needles?
When will the butler conduct us to the altar?

Coffee and biscuits will be served in the patio.
But first observe the Virgin on the walls
Figured in the physiognomy of every nation.
As for example
This elongated Japanese madonna in kimono
With an infant Jesus squatting on the air
At the level of her traditionally obliterated breasts.
They have the sly look of a city whore and her
Dwarfish ponce. But the nations must have their say.

This is no place for art
This might be the place for a cocktail bar
This is no place for little children.

Who shall roll away these stones?
Can any good thing come out of Nazareth?

Give me Manger Square
With its souvenirs, its Flowers from the Holy Land
That I found in my first bible.
Give me the gnarled and bashful Arab
Looking for something shabby enough to kiss.
Give me the disconsolate hippies
Competing for lifts with military heroes.
Or give me the Japanese lady
(For with God nothing shall be impossible)
Come down from the wall and telling me:
'This is no place for love, but I know
A handy cowshed.'

I blame this basilica for my sinful thoughts
This is no place for me.
The reproved couple, their empty arms
Dangling at their sides,
Have been married for quite some time—
If it makes any difference.

Along the River

They had pulled her out of the river. She was dead,
Lying against the rhododendrons sewn with spider's thread.
An oldish woman, in a shabby dress, a straggling stocking,
A worn, despairing face. How could the old do such a thing?

Now forty years have passed. Again I recall that poor
Thing laid along the River Leam, and I look once more.

They have pulled her out of the river. She was dead,
Lying against the rhododendrons sewn with spider's thread.
A youngish woman, in a sodden dress, a straggling stocking,
A sad, appealing face. How can the young do such a thing?

Another person one would like to be

Is a 19th-century composer of
Masses for the Dead.
God knows, one has the emotions anyway
One might as well believe in them.

No call to concoct a plot
No need to write the words
No lack of occasion—
There are masses of dead.

Once I wished I were an old Chinese gentleman
Glimpsed in a Chinese restaurant
Amid masses of Chinese relatives—
With the years one's ambitions grow humbler.

From *The Terrible Shears:*
Scenes from a Twenties Childhood

Two Bad Things in Infant School

Learning bad grammar, then getting blamed for it:
Learning Our Father which art in Heaven.

Bowing our heads to a hurried nurse, and
Hearing the nits rattle down on the paper.

And Two Good Things

Listening to Miss Anthony, our lovely Miss,
Charming us dumb with *The Wind in the Willows*.

Dancing Sellinger's Round, and dancing and
Dancing it, and getting it perfect forever.

Jingle Bells

Our presents were hidden on top of the cupboard.
Climbing up, we found a musical box, in the shape
Of a roller, which you pushed along the floor.

This was for our new sister, she was only
A few months old, her name was Valerie.

Just before Christmas (this I know is a memory
For no one ever spoke of it) the baby quietly
Disgorged a lot of blood, and was taken away.

The musical box disappeared too,
As my sister and I noted with mixed feelings.
We were not too old to play with it.

Where Did Dad Spend Christmas?

Christmas was always a bad time.
My father was on country delivery,
And either he broke his wrist when
Swinging the engine, or the van skidded
Into a ditch with him inside it.

(Once he only suffered from shock.
He ran over a large dog—'it lifted the van
Clean off the ground'—but looking back,
He could see nothing on the road but snow.)

The hours passed and there was no sign
Of Dad. 'We'll let you know as soon as
We hear anything,' said the kindly Post Office.
We began to think he was Father Christmas.

Another Christmas

Another Christmas was coming.
Father thought of a way of enriching it for us.
He recalled that on the Somme
He had carried from the battlefield
A wounded officer by the name of Crawford.

(It was now a household name.)

He wrote to the gentleman in question,
Mentioning the not so distant incident
And the coming of Christmas. In return
There came a free packet of assorted biscuits.

They were consumed, and no doubt enjoyed,
Though felt to be less than they might have been.

Early Discovery

My sense of the superiority of women
Was confirmed at the age of seven.

My young sister was leaning against
The cast-iron railings of the balcony of
The third-floor room in which we lived.
The railings began to fall into the street,
The child began to follow.

My father and I were transfixed.
My mother, though hampered by a bread knife in the right
   hand,
Flew out and pulled the child back with the left.

Then my father had to go down and apologize
To a man in the street whom the cast iron had just missed.

Anglo-Irish

My father claimed to be descended from a king
Called Brian Boru, an ancient hero of Ireland.

My mother said that all Irishmen claimed descent
From kings but the truth was they were Catholics.

We would have preferred to believe our father.
Experience had taught us to trust our mother.

Jephson Gardens

Two small children in the Gardens on Sunday,
Playing quietly at husband and wife.

How sweet, says an old lady, as she sits on
The bench: you must surely be brother and sister?

No, says the boy, we are husband and wife.
How sweet, says the old lady: but really you are
Brother and sister, aren't you now, really?

No, says the boy, trapped in his fantasy,
I am the husband, she is the wife.

The old lady moves off, she doesn't like liars,
She says. She doesn't think we are sweet any longer.

They: Early Horror Film

Our pipes froze last winter
Because THEY took all the heat out of the air.

Not but what THEY are more to be avoided
Than envied.

Remember what THEY did to Joe Walters down the road—
Mucked him about properly.

THEY put Tom Binns in prison for stealing registered letters,
THEY drove up in their Sunbeams and caught him in the act.

THEY're a funny lot, but clever with it.
THEIR maker gave them fancies and the means to gratify
    them.

THEY drink blood after saying grace in Latin.

That Mrs Tooms opposite is stuck-up: she sits at
Her window drinking blood from a teacup, but I bet
She doesn't know any Latin.
Latin isn't for the likes of Mrs Tooms,
Nor is blood.
Even if her son is doing well in the Town Clerk's at
   Folkestone,
Or so she says.

Sometimes THEY try to get round you,
THEY come bearing scholarships.

Keep out of THEIR way, child!
Nothing but shame and sorrow follow.

The divel! Here THEY are, at the door—
Don't open it till I've put something decent on.

'Yes sir, no sir,
We wouldn't know anything about that, sir.'

THEY didn't get anything out of *me*!

Shades of the Prison House

How many remember that nightmare word
The Workhouse? It was like a black canal
Running through our lives.

'Old Mrs Povey has gone to the Workhouse.'
'You'll end in the Workhouse if you go on like that.'

It was shameful to end in the Workhouse.
Shameful to have a relative in the Workhouse,
The worst shame of all.
Such shame was always possible.

Even children came to dread the Workhouse.
Other times, other bogymen.

Young Criminals

'The same to you with knobs on,' said the first.
'The same to you with spikes on,' said the second.
'The same to you with balls on,' said the third.

The master heard us. He was a just man.
The first of us got one stroke on each hand.
I got two strokes on each hand.
The third got a fearful beating—
He was a dunce, he smelt, his name was Bugg.

(About the time I left Clapham Terrace Elementary
He was put away for showing his thing in the street.)

Sunday

My mother's strongest religious feeling
Was that Catholics were a sinister lot;
She would hardly trust even a lapsed one.
My father was a lapsed Catholic.

Yet we were sent to Sunday school.
Perhaps in the spirit that others
Were sent to public schools. It
Might come in useful later on.

In Sunday school a sickly adult
Taught the teachings of a sickly lamb
To a gathering of sickly children.

It was a far cry from that brisk person
Who created the heaven and the earth in
Six days and then took Sunday off.

The churches were run by a picked crew
Of corny actors radiating insincerity.
Not that one thought of them in that way,
One merely disliked the sound of their voices.
I cannot recall one elevated moment in church,
Though as a choirboy I pulled in a useful
Sixpence per month.

Strange, that a sense of religion should
Somehow survive all this grim buffoonery!
Perhaps that brisk old person does exist,
And we are living through his Sunday.

Uncertainties

Our folk didn't have much
In the way of lore.

But I remember a story,
A warning against envy
And also against good fortune,
Too much for our small heads—

About a lucky man called Jim
(My uncle in Dublin I used to think,
But he was Sunny Jim)
And his friend who envied him.
Jim had the luck
He married the girl they both of them loved
And his friend envied him.
Then Jim died, and the friend
Married his widow. And then
The friend envied lucky Jim,
Asleep in peace in the churchyard.

When Granpa wasn't pushing old ladies
Through the streets of the Spa
He would cut the grass on selected graves.
Sometimes we went with him. Dogs
Had done their business on the hummocks.
The water smelt bad in the rusty vases.
The terrible shears went clack clack.

It was too much for our small heads.
Who was it that we mustn't envy—
The living or the dead?

Whatever Sex Was

It was the two sisters next door
Quarrelling over their husband, or
Their drunken husband punching them.

It was the trouble that some woman
Was in, a mysterious trouble that
Could only be talked of in whispers.

It was the man who frightened my
Little sister, and whom my father
Searched the streets for, for hours.

Or the man who got angry with me
In a public lavatory, and followed me
Into the street with inexplicable curses.

It was men fighting outside the Palais.
It was crying, or it was silence.
Whatever sex was, it was another enemy.

A Grand Night

When the film *Tell England* came
To Leamington, my father said,
'That's about Gallipoli—I was there.
I'll call and see the manager . . .'

Before the first showing, the manager
Announced that 'a local resident . . .' etc.
And there was my father on the stage
With a message to the troops from Sir Somebody
Exhorting, condoling or congratulating.
But he was shy, so the manager
Read it out, while he fidgeted.
Then the lights went off, and I thought
I'd lost my father.
The Expedition's casualty rate was 50%.

But it was a grand night,
With free tickets for the two of us.

Euphemisms

After the main entry in clinical Latin and Greek
Which I got by heart in order to dazzle my schoolmates,
The Certificate abruptly changed its tone and remarked
That a Contributory Cause of Death was Septic Teeth.

The oddest thing, however, was to find that the Deceased
Was known as George Roderick. Perhaps this was clinical
Language too. No one ever called him anything but Mick.

Insurance

One spot of cheer in the Midlands gloom
Was our Dublin uncle
Who sent us shamrock each St Patrick's Day
And ebullient letters
On paper headed THE PHOENIX ASSURANCE COMPANY LTD.
He was the family success
(His photograph looks like Mickey Rooney
He could sign his name in careful Gaelic)
He moved in the corridors of power.

He came across for his brother's funeral
Pensively noting the widow and orphans at the graveside.
Something had to be said about our insurance.

He had borrowed, he intimated, the company's notepaper.
Faith, he worked there, he was a janitor there
He moved in the corridors.

Romantic Ireland was dead and gone.

Geriatrics

I got on well with Granma.
Ours was the prescriptive relationship:
We used to play crib for hours, she gave me sweets,
She defended me, her I failed to defend.

It didn't worry me that she was getting
Troublesome. If she wandered in her mind a little,
So did I. Her husband was dead,
So was my father. The house had to be vacated.

We couldn't look after her, we were going
To look after a troublesome old man
Who at least had a house we could live in.
My eldest aunt declined to take her.

She would have to go to the Workhouse.
The worst thing was, they told her
She was going to a nursing home for a while.
They even ordered a car.

She had to be pushed into it.
As the car was moving off, I heard her
Shout with a dreadful new voice:
'I know where you're sending me,
You're sending me to the Workhouse!'

She was found to be deranged on arrival,
And they sent her on to another place.
So she didn't go to the Workhouse after all.
She died soon after.

Early Therapy

Granma doddered a bit.
But she was my friend.
Perhaps it had to be done,
Did it have to be done like that?

It started me writing poems,
Unpleasant and enigmatic,
Which quite rightly no one liked,
But were thought to be 'modern'.

A Much Later Conversation

'Your father—
Did he die in error?'

Well, I suppose you could say that.
He was laid up for several months,
Then they said he was well, and
He went back to work. The next year
He was ill again, and this time he died—

'Yes, but what I mean is—'

Well, a widow's pension wasn't much.
We put it to them that getting gassed
In France had buggered up his lungs.
But the War Office wouldn't wear it,
They observed that the war had been over
For some time . . .

'I'm sorry, but I didn't mean—'

Oh yes, we had a priest in,
The first we'd ever seen at close quarters.
He gabbled in Latin,
And no one could understand him.
But at least—

'I'm sorry, but you've misunderstood me.
I was only asking
Did your father die in Eire?'

No, he died in England.
But you're right, it may have been a mistake.

Learning to Hick and to Hack

'Much have I travelled in the realms of gold
for which I thank the Paddington and Westminster
Public Libraries . . .'—Peter Porter

Behind official bricks
We found the township's casual treasure.

The Public Library
Handsomely stocked and not used to excess
By the public—it was almost like
Having a library of one's own.
Without it, some of us wouldn't have lasted
Much past adolescence.
Into what strange routes it led us,
What pungent semi-understandings!

Bravo, England of the Thirties!
Your smallest dullest town enclosed alternatives
To littleness and dullness.

Only once were we betrayed—
In the Reference Section, where we signed our names
For the Loeb crib of the *Fasti* or the *Civil Wars*.
Loeb was more complete than our school editions,
And our faultless versions gave the game away.
So did our signatures
When the teacher called on the librarian.
I went on rather liking Ovid, though.

Ugly Head

There seems to be a large gap
Somewhere about here.
If repression is at work
Then repression works efficiently,
In this sphere.

I don't remember learning about sex
In the school lavatories;
Though I remember the lavatories.

With a great effort I call up
Certain goings-on in the rear rows
Of the Physics class. I can't believe it.
That Welsh master was so sharp
You couldn't blow your nose
Without him glaring.

At one time or another
Some slightly special one or other—
But to kiss a girl
Would have seemed like criminal assault.
There was one called Pearl
Who would quote bits from Rosalind
In *As You Like It*, leaving me confused.
Once at a party I stepped heavily
On her hand, and was appalled.
In a strange way she seemed to like it.
I was glad to go home and study *The Prelude*.

It was homework and rugger; then
It was essays and walks to Grantchester.
Perhaps we were great Platonic lovers then.
Perhaps there is nothing to remember.

Said the Straw

'It's the last camel,' said the straw,
'The last camel that breaks our backs.'

'Consider the lilies,' he said,
'How they toil in the fields.
Six days they labour,
On Sundays they wear their best clothes.'

'Ask why the violet sickened,
The pale primrose died unmarried,
And the daisy lies in chains,'
He said, 'All grass is flesh.'

'But let us hear no more,'
So said the straw,
'About the sorrows of the camel
With its huge and heavy feet.'

Since Then

So many new crimes since then!—

   simple simony
   manducation of corpses
   infringement of copyright
   offences against the sumptuary laws
   postlapsarian undress
   violation of the Hay-Pauncefote Treaty
   extinction of the dodo
   champerty and malversation
   travelling by public transport without a ticket
   hypergamy and other unnatural practices
   courting in bed
   free verse
   wilful longevity
   dumb insolence
   bootlegging and hijacking
   jackbooting and highlegging
   arsenic and old lace
   robbing a hen-roost
   leaving unattended bombs in unauthorized places
   high dudgeon
   the cod war
   massacre of innocents
   bed-wetting
   escapism
   transporting bibles without a licence

But so many new punishments, too!—

   blinding with science
   death by haranguing
   licking of envelopes
   palpation of the obvious
   invasion of the privacies
   fistula in ano
   hard labour down the minds
   solitary conjunction
   mortification of the self-esteem
   the Plastic Maiden

hat-rack and trouser-press
the death of a thousand budgerigars
spontaneous combustion
self-employment
retooling of the economy
jacks in orifice
boredom of the genitals
trampling by white elephants
deprivation of forgetfulness
loss of pen-finger
severance pay
strap-hanging
early closing
sequestration of the funny-bone
mortgage and deadlock

'Fair do's,' murmured the old Adam, 'I am well pleased.'
He had come a long way since he named the animals.

The Stations of King's Cross

He is seized and bound by the turnstile.

The moving stair writes once, and having writ,
Moves on.

At Hammersmith the nails
At Green Park the tree.

A despatch case which is well named
A square basket made of rattan
Which is a scourge.
The heel of the Serpent bruises Man's instep.

At Earl's Court a Chopper
New, flashing, spotless
It carries hooks and claws and edges
Which wound.

It is hot. Vee
Wipes her face. Cheek to jowl
She wipes the man's on either side.

Rather bear those pills I have
Than fly to others that I know not of.

He speaks to the maidenforms of Jerusalem
Blessed are the paps which never gave suck.

The agony in Covent Garden
He finds them sleeping, for their eyes are heavy.

The first fall, the second fall
The third fall.
And more to come.

A sleeve goes, a leg is torn
A hem is ripped.
This is the parting of garments.

They mock him, offering him vodka.
The effect is shattering.

He is taken down from the strap.
And deposited.

Wilt thou leave him in the loathsome grave?

In Cemeteries

This world a vale of soul-making—
To what intent the finished wares?

Is the ore enforced and fired through
Harsh mills, only to fall aside?

Who is this soulmaster? What say
Do souls have in their made futures?

We mourn the untried young, unmade
In small coffins. What of grown graves?

At times in cemeteries, you hear
Their voices, sad and even-toned,

Almost see the made souls, in their
Curious glory. If you are old.

R-and-R Centre:
An Incident from the Vietnam War

We built a palace for them, made of bedrooms.
We even tracked down playmates for them
(No easy job since prostitutes went out
When self-rule came). We dug a pool,
Constructed shops, and a hut for movies
With benches outside for the girls to wait on.
Serene House was what we called it.
We did our bit in that war.

Air America brought them from the battlefield.
We lifted the girls from the suburbs by buses:
Chinese, Indian, Malay, Eurasian,
Healthy and well-fed and full of play.

There were cameras in plenty, tape-recorders
And binoculars for the soldiers to buy
For the girls; for the girls to sell back
To the shops; for the shops to sell to the soldiers.

Serene House was near the varsity. The GIs
Strayed across the campus with Nikons and blank faces:
It was feared they might assault the female students.
They seemed scared of their own cameras.
They looked at nobody; nobody looked at them.
                              That violence down the road—
It was good for business, and we did our bit.
Otherwise it was a vulgar subject.

Once I found a GI in the corridor,
Young and dazed, gazing at the notice-boards.
The Misses Menon, Lee, Fernandez, Poh and Noor
Should report for a tutorial at 3 p.m.
Bringing their copies of The Revenger's Tragedy . . .
If Mr Sharma fails to pass his essay up this week
He will find himself in serious trouble . . .
The Literary Society seeks help in cutting sandwiches . . .

He was still there thirty minutes later,
A stunned calf. I asked if I could help.
He shrank away: 'Is it not allowed to stand here?'
The corridor was dingy, walls streaked with bat shit,
Somewhere a typewriter clacked like small arms.
'Is there . . . would there be a . . . library?'
One of the best in fact in South-East Asia.—
I offered to show him. He trembled
With a furtive pleasure. His only licence
Was to kill, to copulate and purchase cameras.

What sort of books would he like to see?
Outside in the quad he was jumpy,
As if unused to the open. He glanced behind,
Then whispered. Yes, there was something . . .
Did I think . . .

What could he be after? The Natural History
Of the Poontang, with Plates, by some defrocked
Medico called Aristotle? How to Get to Sweden
By Kon-Tiki through the Indian Ocean?

'Would they have anything . . .' A quick look
Round— '. . . by Cardinal Newman, do you think?'
I left him in the stacks, the *Apologia* in his hands,
He didn't notice when I went away.

Inside Serene House, in the meantime,
Girls galore (such lengths we went to!)
Lolled on the benches, played with binoculars,
Clicked their empty cameras, and groused.
The soldiers were happy to quit Vietnam;
Five days with us, and they were glad to go back,
Rest and recreation, they said, was too much for them.
We weren't surprised when the Americans didn't win.

Home and Colonial

Henri Rousseau's 'Tropical Storm with Tiger'

I'm not one of those simpletons who believe
That if only they had a larger TV screen
They would be able to see the naughty bits.
But if that picture were a few inches longer,
Here on the right-hand side, I mean—then
In fact you would see—not a naughty bit—
You would see me.

Sexual behaviour does exist in the tropics—
Oh indeed—but it's relatively invisible.
It doesn't go on in public. And it wouldn't
Even if there weren't a storm, even if
The jungle weren't so full of spiky things.

Public sex is less sex than public, I reckon.
Like that young couple in the Underground
The other night. They weren't doing anything,
They were simulating it. In my day
We used to dissimulate. And likewise I doubt
This notion that a wider screen creates
A broader mind. What you can see is never
The interesting part. Though of course
I'm not referring to a gentleman like you
Looking at a picture like this
In a reputable gallery.

Imagination is allowed some latitude,
I know (though, as it happens, this painting
Doesn't get enough), but all the same . . .
The jungle's not half as pretty as it looks here,
Untidy at the best, storm or no storm.
The bougainvillaea was tatty and blotched,
Not right out of a hothouse. It was gloomy—
That's another thing about jungles—and
The lightning had that lost air it always has
In those parts. Fumbling around for something
To get a grip on, like a roof, a chimney
Or a golf-club.

But the tiger—Frenchy's hit it off to a T!
Scared stiff, what with its tail behind, which it
Took for a flying snake, and in front—a hairy
Red-faced white man in a post-impressionist sarong,
Heading for the nearest drinking shop.
I fancied an ice-cold Guinness. A moment later
And there'd have been just me on that canvas,
Dry and wet at once, sarong slipping a bit,
Tiger a mile away and still running.

Even so, would you really see more on a larger screen,
D'you think? Or do the girls wear towels or something?

Very Clean Old Man

Do you know that land?—
Where the nocturnal tiger
Empties the rubbish bins,

Where the deathless mosquito
Stings the flesh to life
And the body runs with oils!

This splendid desk of teak
I brought from that lost land,
Large, larger than my room—

Now in this temperate clime
Cracks have run across it,
Large, larger every day.

My nails slipped through them,
Then my fingers—next
My typewriter . . .

If you change your country
Change while you are young,
Before your bones grow brittle
And your life cracks across.

These suburbs of the West,
Dog-haunted, dream
Of soap powders, and the wind
Empties the rubbish bins . . .

Maybe you should try the moon
—You cracked old man—
Once they dust it down.

Oyster Lament

We are a poor people, who
Cannot afford oysters any more.
The sea is a long way away
Nowadays.

Freedom we have in plenty,
Golden and tall it waves in the fields.
How lovingly somebody tilled the soil,
Manured it so richly!

But who can eat that much? What happened
To those overseas markets? Something is wrong
With the rate of exchange.

It makes such a rattle against the panes.
It is cracking the tarmac out in the streets.
They say it uses up the oxygen.

Freedom is a pearl, to be sure,
A pearl above price.
But so are oysters.
Often I think I would rather have oysters,
Their taste, their indefinable taste.

Buy One Now

This is a new sort of Poem,
It is Biological.
It contains a special Ingredient
(Pat. pend.) which makes it different
From other brands of poem on the market.

This new Poem does the work for you.
Just drop your mind into it
And leave it to soak
While you relax with the telly
Or go out to the pub
Or (if that is what you like)
You read a novel.

It does the work for you
While (if that is what you like)
You sleep. For it is Biological
(Pat. pend.), it penetrates
Into the darkest recesses,
It removes the understains
Which it is difficult for us
Even to speak of.

Its action is so gentle
That the most delicate mind is unharmed.
This new sort of Poem
Contains an exclusive new Ingredient
(Known only to every jackass in the trade)
And can be found in practically any magazine
You care to mention.

From *Paradise Illustrated: A Sequence*

I

'Come!' spoke the Almighty to Adam.
'There's work to do, even in Eden.'

'I want to see what you'll call them,'
The Lord said. 'It's a good day for it.'
'And take your thumb out of your mouth,'
He added. (Adam was missing his mother.)

So they shuffled past, or they hopped,
Or they waddled. The beasts of the field
And the fowls of the air,
Pretending not to notice him.

'Speak up now,' said the Lord God briskly.
'Give each and every one the name thereof.'

'Fido,' said Adam, thinking hard,
As the animals went past him one by one,
'Bambi', 'Harpy', 'Pooh',
'Incitatus', 'Acidosis', 'Apparat',
'Krafft-Ebing', 'Indo-China', 'Schnorkel',
'Buggins', 'Bollock'—

'Bullock will do,' said the Lord God, 'I like it.
The rest are rubbish. You must try again tomorrow.'

II

'What a dream!' said Adam waking.
'I never dreamt a dream like that before.'

'You will,' remarked the Voice Divine, 'You will.'
'And worse,' He added *sotto Voce*,
Finding it hard to speak in accents mild,
Knowing what He knew.

('Why can't I ever live in the present?'
He would grumble. 'Never in the present.')

100

'You're luckier than I,' the Almighty said.
'I know of no one fit to shake My hand,
Let alone My equal. I'm on My own.'

'You're different,' said Adam,
'You don't need it.'
'It's wearing off,' said Adam,
'The anaesthetic.'

'Anaesthetic? What's a man like you
To do with words like that?'
He found it hard to speak in accents mild,
Knowing what He knew.

'You promised, Lord,'
Urged Adam. 'You promised me.'

'Behold her, not far off,
Flesh of your flesh, bone of your bone,'
Said He in neutral tones,
'Your madam, Adam,'
Knowing what He knew.

     III
'Rich soil,' remarked the Landlord.
'Lavishly watered.' Streams to the right,
Fountains to the left.
'The rose, you observe, is without a thorn.'

'What's a thorn?' asked Adam.
'Something you have in your side,'
The Landlord replied.

'And since there are no seasons
All the flowers bloom all the time.'

'What's a season?' Eve inquired.
'Yours not to reason why,'
The Landlord made reply.

Odours rose from the trees,
Grapes fell from the vines,
The sand was made of gold,
The pebbles were made of pearls.

'I've never seen the like,' said Eve.
'Naturally,' the Landlord smiled.

'It's unimaginable!' sighed Adam.
'You're not obliged to imagine it,'
Snapped the Landlord. 'Yet.'

IV

If you wanted ice-cream
There was ice-cream galore
Oozing from handy ice-plants.

(But you didn't really want ice-cream:
The weather wasn't hot enough.)

If you wanted a piping hot bath
There was piping hot water
Running in convenient brooks.

(But you didn't really want a piping hot bath:
The weather wasn't cold enough.)

If you wanted petrol
There was petrol in plenty
A few inches under the Garden.

(But what was the need for petrol?
There was nowhere you wanted to go.)

If you wanted money
Money grew on trees
(But what would you do with money?).

'But Adam wants me,' Eve told herself.
Adam told himself, 'But Eve needs me.'

V
'About them frisking played
All beasts of th'earth . . .'

'If we have a baby,
That elephant will have to go—
He's too unwieldy.'

'What's a baby?'
'A word I've just made up,'
Said Adam smugly.

'If we have a child,
That bear will have to go—
He's wild.'

'What's a child?'
'A word I've just made up,'
He smiled.

'If we have a brood,
That ape will have to go—
He's very rude.'

'What's a brood?'
'A word I've just made up,'
He cooed.

She said:
'But won't the wolf lie down with the lamb?'
He said:
'I think of words, therefore I am.'

VI
'Can't you let *her* name something?'
Begged Adam. 'She's always on at me
About the animals.'

'Herself a fairer flower,'
Murmured God. 'Hardly necessary,
I would say. But if it makes her happy . . .'

*

'What a trek!' Eve muttered.
'The animals *came* to Adam . . .
Well, Mohammed must go to the mountain.'

'What's that you said?' the Almighty asked.
But she was on her way.

*

'Lady's finger,' said Eve.
'Lady's smock.
Lady's slipper.
Lady's tresses . . .'

She paused.
'Adam's apple.'

'No,' said the Lord,
'Strike that out.'

'Old man's beard, then.'
She sped towards the mountain.

'Lily.
Rose.
Violet.
Daisy.
Poppy.
Amaryllis.
Eglantine.
Veronica.
Marigold.
Iris.
Marguerite.
Pansy.
Petunia.
Jasmine.
May.'

'I'm worn out,' she gasped.
'Belladonna—
And that's all for today.'

*

'She's better at names than you were,'
The Lord observed.
'They all sound womanish to me,'
Said Adam, nettled.

VIII
'I thought I had no family ties,'
Thus the Ruined Angel sighs.
'But now I find I have a daughter.
By her I find I have a son.
I find my son has children by his mother
(Such sad dogs too).

That makes some family!
Sin and Death and all their progeny.
Not counting Lilith
Whom, truth to tell,
I can't remember well.

Eden's the right place for a family—
Where life is new and easy,
With first-grade fruits from fertile trees.
There shall they be fed
And filled immeasurably.

The present population
Consists of two small upstart squatters,
Childless, as I chance to know.
The greater good of the greater number . . .
Those two will have to go.'

IX
Satan considered the creatures.
Satan selected the serpent.

'The subtle snake, the fittest imp of fraud,'
So spoke the Fallen Angel, fond of artful sound,
So spoke the Fiend, alliteration's friend.

Unfeared and unafraid,
The silken snake lay sleeping . . .

In at his mouth the Enemy entered,
Thorough his throat commodious
Filtered our fatal Foe.

Then up he rose—
As yet the snake stood upright on his tail,
This sleek unfallen fellow,
He surged, not slithered—

And off the Tempter tripped,
And off swept Satan Snakeskin,
In search of silly She.

      X
'All this fuss about an apple!
Now, you're a simple woman,'

Said the Snake. 'Like me.
I mean, I'm a simple serpent.

It's a conspiracy of course.
It's meant to keep you down.'

'I don't mind being down,'
Eve tittered.

'Tut-tut,' the Snake tut-tutted.
'Think of higher things!

He only wants to keep you dumb,
And that's the truth.

What's wrong with *knowing*?
Do *you* know what?'

'What's good for talking snakes
Is good for me,' she mumbled,

'Pity to leave it for the birds.'
But the Snake had scarpered.

XI
Eve chomped at the apple
(Her teeth were white and strong:
There was no such thing as decay).
The juice ran down her chin.

She ate it all
(There was no worm at the core:
There was no core).
The juice ran down her breasts.

'I have done something original,'
She told herself.
'But I mustn't be selfish.'
She plucked a second for Adam.

XII
'What about Adam?
Shall I keep it to myself,
Shall I be on top?

For who, inferior, is free?

But what if I drop dead?
He rushes off and marries someone else.

Good or bad,
Dead or alive,
On top or underneath,
I'd better share it.

I'm a simple woman.'

XIII
Sighing through all her works,
Nature gave signs of woe.
Earth trembled from her entrails,
Nature gave a second groan.

*

'What's that strange noise?' asked Eve.

'Nothing to worry about,' said Adam.
'Just cataclysms, convulsions, calamities—'

'Don't talk with your mouth full,' said Eve.

'Donner-und-Blitzen, coups-de-foudre, infernos,
Avalanches, defoliation, earthquakes, eruptions,
Tempests, turbulence, typhoons and torrents,'
Said Adam airily.

'And floods. Or do I mean droughts?'
He pondered. 'Also perhaps inclemency.'

'The Snake was right about one thing,'
Eve observed. 'It loosens the tongue.'

XIV
'Must be that Cox's orange pippin,'
Said Adam. 'I do feel queer.'

'I can see that,' said Eve.
'Come over here.'

'Here?'
'Here—

You're good at naming things.
What would you call this?'

Whisper, whisper,
Titter, titter.

'And this—
What would you call this pretty thing?'

Mumble, mumble,
Giggle, giggle.

'Hey, you never did that before!'
'It must be what they mean by sex-apple.'

'Ouch.'
'Mmm.'

'I wish there were ten forbidden trees.'
Snore.

XV
'Where are you, Adam?'

'I'm behind this tree, Lord.'

'What are you doing there?'

'Nothing, Lord. I'm naked is all.'

'Who told you you were naked?'

'I noticed it, Lord.'

'Naked, shmaked, what does it matter?'

'You can get six months for indecency, Lord.'

'What's six months to you, you're immortal.'

'It might give the animals funny thoughts.'

'My animals don't have funny thoughts.'

'The best people wear suits, Lord.'

'You were the best people, Adam.'

'The weather might change, Lord.'

'Too right. It will.'

XIX

So the Archangel, out of pity,
Now disclosed the liberal arts
That should relieve man's fallen lot.
'Like music, painting, plays and books.'

'Long books?' asked doleful Adam,
Whom the stern Angel had apprised
Of death and rape and guns and hunger.

'A book there'll be,' the Angel said,
'About this very business—
A poem of ten thousand lines, which one
Called Milton shall compose in time to come.'

'Oh dear!' Then Adam brightened.
'Am I the hero of this book perchance?'
'Not quite the hero,' Michael mildly said,
'And yet you feature largely in it—
God, not unnaturally, is the hero.'
'Should have known,' groaned Adam.

'Although there are—or will be—those
Who claim the hero really is—or will be—
Satan. As I of late foretold,
Henceforth the human race is fallible.'

'That circus snake?' hissed Adam scornfully.
Eve hid her blushes in her work,
A garment she was knitting, made with
Real lamb's wool, tight-fitting.

'In my opinion, which I trust
You won't repeat,' the Angel whispered,
'The hero really is the Son,
Called Jesus, even though his lines
Are fewer in the poem than are mine.'

'And me?' Eve raised her eyes. 'Am I in this—
This book of yours? Or, as I well suppose,
Are all the characters men?'

'Indeed you are!' the genial Angel cried,
'Without an Eve there'd be no tale.
While Mr Milton's not a woman's man,
He does your beauty justice, and your brains.'

'A female intellectual?' Eve grew vexed,
Old-fashioned in her ways as yet.
'No,' spoke the nervous Angel, blushing more,
'I only meant, not just a pretty face.'

Eve held the knitting to her breast.
'By me the Promised Seed shall all restore.'
And Michael knew the time was ripe to leave.
'All—or some,' he murmured at the door.

     XX
There was the escort, in position,
Presenting fiery arms and flaming brands—
Handpicked Cherubim, four faces
To each man and pairs of eyes to match.

'A splendid turn-out!' Eve declared,
Waving graciously.
'Not like our coming in,' said Adam,
Grabbing her wrist.

'Hurry up please!' said the Archangel.
His men were sweating, the Sword of God
Had started the stubble burning.
'We don't want any trouble, do we?'

     XXI
'I had no voice,'
Sighed Adam. 'No real choice.'

Eve wiped a tear and said,
'I wanted you, alive or dead.'

'Whether boon or curse.'
'For better or for worse.'

'Come fair or foul, we have to eat'—
She watched for herbs beneath their feet.

The world lay there, and they could choose.
He said, 'I'll learn to make you shoes.'

XXII
'Why didn't we think of clothes before?'
Asked Adam,
Removing Eve's.

'Why did we ever think of clothes?'
Asked Eve,
Laundering Adam's.

XXVI
'What's all this
About a sense of sin?
It's rot!' cried young Cain,
'It's the opium of the people.'

'It's all we have, son,'
Said Adam. 'We grew up with it.'
'What's opium?'
Asked Eve suspiciously.

Later Cain killed his brother
For business reasons.
'You see?' his parents said,
'We told you so.'

XXIX

There were giants in the earth.
Some said they were angels
Lusting after pretty Jewish girls.
Others said they were scholars
(That is, 'mental giants')
Lusting after pretty Jewish girls.

Either way, their offspring turned out ill,
They went to the Devil.
Who had been living in partial retirement,
Teaching himself German:
'Ich bin der Geist der stets verneint!—
Oh very well then, I accept.'

'I knew this would happen,' said God.

XXXI

'Cities were all right when they were towns,'
Adam grumbled.
'Now it's all rush and noise and pollution.'

'Only this morning I stepped in something
Nasty,' said Eve. 'Left by a dog.
It shouldn't be permitted.'

'Don't blame me then,' said Adam.
'Dogs were always like that.'
Give a dog a bad name, he thought.

'And horses are no better. Look—
Just in front of the house!
You'd wonder where it all comes from.'

Adam rushed for a shovel.
'Can't have you stepping in it, dear.'
It was just the thing for his allotment.

XXXII

The days of Adam were 930 years.
He sat in the market-place
With other senior citizens.

With Seth, who was just 800,
Enos, who was 695,
And Methuselah, only 243.

'They're not the men
Their great-grandfathers were,'
Said Seth.

'Lamech's kid Noah cries all night,'
Said Enos.
'Howls when they bath it.'

Said the youthful Methuselah:
'They've all been spoilt.
I blame their mothers.'

'It was different in my day,'
Said Adam.
'People lived for ever then.'

XXXIII

Adam was old bones, lost in his clothes.

Eve thought of how they met. Long,
Long ago. It must have been on holiday.
Warm the air, and full of scents.
And he was wearing nothing, not a stitch.
She blushed a little, pleasurably.

Adam's dreams were not so pleasing.
He woke, delirious and sweating,
Tearing at his twisted nightshirt.

'Eve . . . my side is bleeding . . .
A spear stuck in my side. My rib is hurting.'
'Look,' she said, 'there's nothing there,
No sign at all, dear. Go to sleep.'

He fell back, mumbling indistinctly
About a seed, a promised seed.
She wiped his brow, and smiled:
At heart he'd always been a gardener.

   XXXIV
'The Lord called Adam *Man*.
But Adam called me *Life* itself.

Adam is dead now, man is not.
The Man is dead, but long live man.
For Life goes on.

Priests everywhere, police
And politicians, laying down the law.'
The old girl grinned.

'And I it was who broke it . . .
Poor Adam, poor old Man,
Our sin was self-sufficiency he said.'

Eve did the crossword by her lonely fire.
'That which clasps, a tendril: RIB.'
She died soon after Adam.

The Verb 'To Think'

The verb 'to think'
Is represented by the same sign as
'To yearn for', 'to be sad',
'To be unable to forget'

In many cases
'To love' is rendered as 'to like'
(Thus, a man is said to 'like' his wife)

The symbol for 'feeling' also signifies
'A certificate of merit'.
It is meritorious to feel,
Though 'to love' is more often 'to like'

For love comes in two forms alone:
That of a mother for her child
And carnal

The verb 'to do' is homophonous with
The verb 'to pick someone's pocket'.
This, we should note, is not
(As it is in some languages) slang

Slang would be to say that you 'love'
Someone or something when more properly
You 'like'

The verb 'to be'
Seems not to exist.
Yet you cannot say that the verb 'to be'
Does not seem 'to be'

'To be sad' does not imply 'to be',
It implies 'to think'.
'To think' is 'to yearn for', 'to be without',
And to be a man of feeling is to be certified.
'To like' is more common than 'to love'
And 'to be' is most uncommon—

Unwise it is, in the long watches,
When the leaves litter the misty streets,
To be reading a Japanese lexicon
And to be unable to forget.

The Answer

It was with aesthetic satisfaction and
Intense moral relief that he began at last
To see a figure forming in the carpet.

'Excuse me sir,' the carpet murmured faintly,
'I am gratified that you perceive a pattern,
But people continue to walk all over me.'

And there it was, the figure in the carpet!

The Word

The sage said: We are all books
In the great Library of God.
(He was a bookish person.)

One asked: Does He ever
Take us out?
We spend our years as a tale that is told.

The sage said: His will be done
In the Library as it is Elsewhere.

One asked: But perhaps
He is only interested in first editions,
Not in reprints, abridgements, strip cartoon
Or other adaptations?

The sage said: His love speaks volumes.
He is a speed reader. He is no respecter
Of Bestseller lists.
He suffers the little magazines to come unto Him.

Some hoped their jackets would be clean
And well pressed when the call was heard,
Their loins girded about and their lights burning.

God thought: I wrote all the books,
Now they expect Me to read them.

Seaside Sensation

The strains of an elastic band
Waft softly o'er the sandy strand.
The maestro stretches out his hands
To bless the bandiest of bands.

Their instruments are big and heavy—
A glockenspiel for spieling Glock,
A handsome bandsome cuckoo clock
For use in Strauss (Johann not Lévi),

Deep-throated timpani in rows
For symphonies by Berlioz,
And lutes and flutes and concertinas,
Serpents, shawms and ocarinas.

The sun is shining, there are miles
Of peeling skin and healing smiles.
Also water which is doing
What it ought to, fro- and to-ing.

But can the band the bandstand stand?
Or can the bandstand stand the band?
The sand, the sand, it cannot stand
The strain of bandstand and a band!

Now swallowed up are band and stand
And smiling faces black and tanned.
The sand was quick and they were slow.
You hear them playing on below.

Poem Translated from a Foreign Language

My mother once told me
Never to sleep with flowers in the room.
Since then I have never slept
With flowers.

I have slept with prostitutes
(But never with flowers in the room)
In general they were amiable
A few were not unhumorous
One was rather acid.

None of them smothered me.
One of them desired a silver pen
Another removed a traveller's cheque
But never my precious oxygen.

My mother never told me
Not to sleep with prostitutes in the room
She did not tell me to stay awake.
These days I sleep with pills
I sleep with daffodils in the room.

History of World Languages

They spoke the loveliest of languages.
Their tongues entwined in Persian, ran
And fused. Words kissed, a phrase embraced,
Verbs conjugated sweetly. Verse began.
So Eve and Adam lapped each other up
The livelong day, the lyric night.

Of all known tongues most suasive
Was the Snake's. His oratory was Arabic,
Whose simile and rhetoric seduced her
('Sovran of creatures, universal dame').
So potent its appeal—
The apple asking for eating,
To eat it she was game.

Now Gabriel turned up, the scholars say,
Shouting in Turkish. Harsh and menacing,
But late. And sounds like swords were swung.
Fault was underlined, and crime defined.
The gate slammed with the clangour of his tongue.

Eden was gone. A lot of other things
Were won. Or done. Or suffered.
Thorns and thistles, dust and dearth.
The words were all before them, which to choose.
Their tongues now turned to English,
With its colonies of twangs.
And they were down to earth.

From *A Faust Book*

Dr Faust discourses on the subject
of training and heredity

Out for a healthy walk, Faustus picks
Up a poodle, or vice versa.
A comical little fellow, it chases sticks,
Crawls on its belly, races in circles.

'Perhaps you could train it,' says Wagner.
'No doubt it would respond better
Than many of my students,'
Remarks the Professor.

'Even so,' his assistant ventures,
'There's the question of pedigree.'
'A dog's a dog for all that,'
Says the Professor unctuously.

'To heel, Blackie!' Blackie wags its tail.
'Man's best friend,' says the Professor,
A shade sentimental, it being Easter.
The dog assumes a toothy smile.

Faust in his study

'The interesting questions go unanswered—
Undue suffering, undeserved success . . .

Divinity talks all the time of better times to come,
Science invents false limbs and true explosives,
And Alchemy turns lead to lead at vast expense;
Art thrills with piggish gods and puffy goddesses,
While Law invokes the laws its lawyers legalize,
And Logic says it all depends on what you mean . . .

So must I live and die in Aristotle's works?'

*

'Then what is life,'
Faust paced between his piles of parchments,
'Except a slower form of suicide?'

A first-class speaker,
He found himself a first-class listener too.

'This is indeed an honour,'
Yapped the dog. 'I could crouch here
Listening to you for all eternity,
I mean all day.'

'I thought there was more to you
Than met the eye,' said Faust,
Patting him.

'Talking of suicide,' the dog remarked,
'Perhaps I can be of some small service,'
Patting Faust.

Mephistopheles enlightens Dr Faust

'Why, if I may ask, did you appear
In the shape of a dog, Mephistopheles?'

'As you will know, our liturgies
Run widdershins, dear Doctor—
A backwards Paternoster,
Commending Cain, rejoicing in Judas,
The Black Mass et cetera.
In my view, often a little vulgar—
Part of the hearts and minds manoeuvre
To win over the masses . . .

So in our parlance the Lord God
Is the Drol Dog,
Ergo, a poodle.
Nothing could be more simple . . .

Talking of which, every god has its day—
What would you like to do with yours?'

Mephistopheles desires a few lines in writing

'Being of sound mind (and all that mumbo-jumbo), I (name in full) by these presents (already?) do covenant and grant, in consideration of 24 (Twenty-Four) years of service by Mephistopheles, the which (which?) shall bring me anything, or do for me whatsoever, that I give them (names of Lucifer and self) full power to do with me at their pleasure, to rule, to send, fetch, or carry me (that should cover everything) or mine, be it either body, soul, flesh, blood, or goods (curious expression), into their habitation, be it wheresoever (how discreet!) . . .'

*

'Here are the papers,
All ready for signature—
I'm yours for 24 years,
You're mine thereafter . . .

Now where's my penknife?
Quick, before it dries . . .
Blood is thicker than ink,
And I hate to strain my eyes.

I've immense respect for symbols,
They mean a lot to me.
Noughts and crosses, for example,
And the loaden tree.'

(Some other words were heard in Faustus' mind.
There is in love a sweetnesse readie penn'd:
Copie out onely that, and save expense—
But reason could not tell him what they meant.)

'Anaemic forsooth!
Lucky I'm not that young Count
With a sharp sweet tooth
And a taste for throats . . .

Words appearing on your palm?
Sounds most unlikely!
It looks like what they call a Rorschach Blot,
They say it shows your personality.

124

That's right!—
Now he who runs may read . . .
Oh do stop fussing!
A little spittle clears you of this deed.'

Mephistopheles introduces Faust to the fourth estate

Under the good solid name of Gutenberg Inc.
Mephisto instructs Faust in the art of printing . . .

(The wicked sprites strike up on cue:
    'We have done our best or worst,
    Faust or Fust shall be the first—
    Black letters on white background seen
    Will prove our powder-magazine . . .'
As infantile as ever.)

Whereupon Herr Gutenberg declares himself bankrupt
—'This fairy gold does tend to depreciate'—
And hands the business over to Dr Faust or Fust.

'I have taught you everything I know,
The rest is up to you, my boy—
Take care of your running head, and
Your tailpiece will take care of itself.'

'There aren't many authors about,' Faust complains.
'The Imperial Treasury ought to make grants to writers.'

'I know an author,' says helpful Mephistopheles,
'Whose work is out of copyright moreover.'

*

So Faustus printed the Bible,
In black letters on white background.
And after the copies were all bound
He discovered a word was missing—
'Thou shalt commit adultery.'

'You've turned the Good Book into a naughty one,'
Mephisto laughed. 'I can teach you nothing.'
'You can't get decent proof-readers,'
Moaned Faust, building a bonfire.

He began again, and when the books were distributed
Someone detected a dangerous heresy—
'The fool hath said in his heart there is a God.'
Faust was fined £3000 and the book was suppressed.

'Here's fool's gold to pay a fool's fine,'
Mephisto offered generously.
'You had better specialize in French romances.'

*Faust requires Mephistopheles to describe hell and
heaven for him*

'Tell me, I command you,
What is that place called hell?'

'Hell is grey and has no bottom—
Which reminds me of a joke I heard . . .'

'It was of hell I asked to hear,
Not your stale jokes.'

'But hell is of no substance,
A confused and hence confusing thing—'

'It is yourself that you describe,
I think.'

'So be it—hell is,
Well, it's empty.
Oh there are crowds of people all around.
But hell, you feel, is—empty.
The word is emptiness.'

*

'Now let us speak of heaven.
What manner of place is that?'

'I lack authority to speak of such.'
(Though not to miss it.)

'Yet you were once in heaven,
Or so I've heard.'

'Some time ago, it may have changed . . .
But—heaven is full.
That's not to say it's chock-a-block with people.
Simply, full—that's what you feel.
The word is fullness.'

'For one renowned for eloquence
I find you sadly tongue-tied!'

'Whereof one cannot bear to speak,
Thereof one says but little.'

The Doctor displays his erudition in philology

'Do you mind if we go by broomstick?'
Mephistopheles asked.
'It wouldn't look right, arriving by cab.'

       *

'The word *sabbat* or *sabbath*,' said Faust,
'Does not derive from the mystic number *seven*,
As is commonly believed . . .'

'Number seven—go to heaven,'
Mephisto murmured, busy at the controls.

'An amusing though scarcely tenable theory
Has it that the word comes from *Sabazius*,
A Phrygian deity akin to Bacchus . . .'

'Quite a strong head wind,' the pilot grumbled.
'Steeds would have been steadier.'

'But the consensus of opinion has it that *sabbath*
Comes from the Hebrew *shābath*, meaning *to rest* . . .'

'There's no rest for the wicked.'
The pilot corrected a nose dive.
'New brooms don't always sweep clean.'

'My own view, necessarily tentative,
Is that it stems from *saba*, as in *sabaism*
Or *star worship*, properly *host of heaven* . . .'

'In the vernacular,' Mephisto interjected,
'*Army of devils* . . . Bend your knees, please,
We're coming in to land!'

'Interesting word, *knee*: from Germanic *knewam*,
Cognate with Latin *genu*, whence quite possibly
We take *ingenuous*, inclined to kneel—'

                          *

'I can't see you enjoying the orgy,'
Mephisto scolded, 'with a broken ankle.
For you the word is *shābath* . . .'

Faust is struck by fears for his career

'There's an academic joke,
A funny story, well, meant to be funny . . .'

'Indeed?' says Mephistopheles glumly,
Not very fond of Faustus' funny stories.

'Well, you see,
A soldier squatting at the foot of the Cross
Says to another:
If the fellow's as bright as you say,
Excellent speaker, always helping people . . .
Then why is he *up there*?—'

Mephistopheles starts laughing.
'Not yet,' says Faust testily,
'I haven't got to the point yet . . .

Then the other soldier says:
Ah, no publications!'

'They came later,' Mephistopheles remarks,
Forgetting to laugh.

'It's a *joke!*' Faust snarls,
'About getting on in the world . . .'

'I thought you were what they call
Permanent,' his friend says soothingly.

'But Wittenberg isn't an awfully prestigious
Place . . . I was thinking about Berlin,
The Sorbonne, or even East Anglia . . .

I might produce a book on Metaphysical Evil,
Of course—with some help from you . . .'

'By all means,' his friend replies.
'To coin a phrase—be damned and publish!'

Faust quizzes Mephistopheles about love
'But what can fallen angels know of love?'

'We know desire that nothing will relieve.'

'There surely must be whores enough in hell!'

'And all retired . . . My theme is something else.'

'What do you do with love instead of loving?'

'We sublimate our fruitless longing.'

'So hell is paved with high ideals . . . But how?'

'We take an interest in such as you.'

129

Mephistopheles plays the go-between

'What is love?'
    Mephisto asked Faust cheerily.
''Tis not hereafter.' (None there embrace.)

'A boy-friend is a girl's best friend,'
    he then told Gretchen earnestly.
'And diamonds next.' (Paste in her case.)

These things were quick to fix.
Faust lived in a big house with lots of space,
Gretchen's cottage was dark and poky.
The sexes lent themselves to sex.
He hadn't really needed jewels. Or poetry.

Faust and Gretchen walk in the garden

'No, I don't see much of the priest—
Actually I'm a Doctor of Divinity myself . . .'

If Divinity was sick
He was the man to cure it.

'Do I believe in God, you ask.
You're a natural philosopher!
Can anyone say he believes in God?
What is meant by *God*?
What is meant by *believe*?
What is meant by *I*?
Can we employ these words any more?'

Gretchen felt that she could,
Though in fact she rarely did.

'But when I speak of *hand*
I know what I mean. I can touch it . . .
Ah, let me kiss it!
What do you mean, I don't know
Where it's been?'

It had been in soap and water.
Her mother took in washing.

'You're not very fond of my skinny friend?
I wouldn't want you to be!'

Gretchen shuddered. Her hand moved
To the crucifix round her neck.

'So you're wearing a cross,
You little darling!
I gave it to you?
Yes of course I did.'

He remembered Mephisto grinning
As he handed it over.
'When in Rome . . .' the fellow had said.

'It's too good for you?
Nonsense, my sweet—nothing's too good . . .
Oh dear, the chain's broken—
Where can the cross have got to?
It can't have gone too far . . .'

It went too far.

A game of consequences

It had rained the whole day long,
    And Faust was moping.

Mephisto said to Faust:
    'Let's have an evening out.'

Faust said to Mephisto:
    'As long as you're paying.'

So they went to Spicy Spiess's
    In downtown Wittenberg.

There Faust met a go-go dancer
    Called Meretrix, big bust, long legs,

131

Lusted after by university wits
  And city aldermen alike.

Faust said:
  'I thirst.'

Mephisto told him:
  'Try this Venusberg Lager.'

Meretrix said: 'I'm crazy about theologians!
  Make me an offer I can't refuse.'

Mephisto did, the handsome devil.
  Consequently, there were consequences.

*Faust desires to have as his paramour that beautifull
and delightfull peece Helen of Troy*

'Sweet Helen, make me immortal with a kiss!'
Mephisto nods: there was no objection to that.

Golden tresses, swanlike neck and coal-black eyes—
But was she bold and wanton as bespoken?

'Thy lips suck forth my soul: see where it flies!'
Mephisto shakes his head: it should stay where it was.

'That's enough of this youthful idealistic stuff,'
He hisses to Helen: 'You know your business.'

She glares coldly at him, not yet finding her tongue.
She thought her function was to classicize.

Mephistopheles and the primrose

Mephisto smiled at the primrose,
The primrose smiled back.

'A primrose by a river's brim,'
Said Mephisto sanctimoniously,
'A yellow primrose is to him—
Faust, that is—
And it is nothing more.'

The primrose looked unsure.
'But what am I to you, sir?
I feel I can talk to you.'

'To me,' he replied judiciously,
'A primrose by a river's brim
Is something which to him—
I refer to Faust—
Is nothing more than a yellow primrose.'

The primrose dropped a tear,
Or perhaps a drop of dew.
'I just can't win, I fear . . .'

'Cheer up,' said Mephisto.
'I am sure there is someone somewhere
Who regards you as something more
Than a yellow primrose.'

'Oh?' The primrose perked up.
'And who might that be?'

'The Lord God,' said Mephisto.
'After all, he made thee.'

Adding thoughtfully, 'Mind you,
He also made a host of golden daffodils,
To mention but a few.'

Faust and the daffodils

'Top of the morning to you!'
Cried jolly Faustus.

The daffodils tossed their heads,
They too were jocund.

'What a crowd of golden daffodils!'
He exclaimed. 'Or even a host . . .'

The daffodils tossed their heads again.

'One might say you were fluttering
And dancing in the breeze . . .'

The daffodils nodded vigorously.

'But mind your heads don't drop off,'
Said Faustus, bending down.
'Only the best will do for Helen.'

*In extenuation of taking a concubine, Faust pleads
vagueness*

'An affair with an immaterial and possibly
Mythical figure hundreds of years old—
How could that get me into serious trouble?
If it were the Rektor's wife I could understand . . .'

'If it were safe there wouldn't be any fun in it,
Would there?' said Mephistopheles.
'Remove the sin from penis and there's not much left,
Is there? You might as well take Origen's tip.'

This is the fellow who calls me trivial,
He thought to himself.
'Do you really fancy the Rektor's wife?' he asked.

*

'Faust was right,' mused Mephistopheles.
'All this sex isn't getting me far,
And he's even enjoying it . . .

A shady lady with a shady past,
A go-go girl paid well and on the nail,
A silly miss called (save us!) Gretchen . . .

If only I could get him into bed
With one of his students!
But that sticks in his gullet . . .'

But then he brightened.
'Love of course is another matter.
Love calls for concentration—

A thing he hasn't time for.
At least, if I've not harmed him much,
I haven't done him good . . .'

Helen vanishes

'But she's fading away,'
Faust said in dismay.
'We must send for a doctor.'

Mephisto shuddered.
That was the trouble with shades,
They faded away.

'I can see right through her!'
Lamented Faust.
'Was this the face . . .?'

Mephisto shrugged.
That was the trouble with cohabitation,
You soon saw through it.

Then one day she wasn't there.
'Gone to Paris for the weekend,'
Mephisto ventured.

But she never returned.

Walking in the Harz Mountains,
Faust senses the presence of God

God was a brooding presence.
Brooding at present over new metres.
In which his creatures could approach him,
In which they could evade him,

—And he be relieved of their presence,
Through art as Proxy Divine—
Sublimation, as they termed it,
Which could very nearly be sublime—
For which he was truly thankful.

But how active they were, the bad ones!
They brooded rarely.
They talked incessantly,
In poisoned prose from pointed tongues.
How gregarious they were!
They needed friends to wound.

But who had invented tongues?
(One had to be careful when one brooded.)
And even the better ones
(One had to remember)
Were only human . . .
He started to fashion a special measure
For the likes of Gretchen, a still, sad music.

Creation was never finished.

Gretchen at the sewing-machine

'My stomach is sore,
My peace is fled,
I wish I'd been careful,
I wish I was wed.

Once I would scold
A poor girl in disgrace,
Now it's for me
To hide my face.

I love him still,
My dear Johannes,
His clever words,
His polished manners.

It's not surprising,
When I love as I do,
That those more knowing
In love, love him too.

There's no one to help,
I can't tell my mother,
The priest would be shocked,
I'm scared of my brother.

He says his religion
Forbids him to marry,
If only I
Had the luck to miscarry!

Perhaps his crony,
That creepy fellow,
Could tell me of someone
Or somewhere to go.

My stomach is swelling,
My peace is fled,
I wish I'd been careful,
I wish I was dead.'

Faust exhibits his necromantic powers

At the Duke of Parma's court
Faust performed unheard-of feats of magic.

He summoned spirits to enact
Goliath slain by David,
Judith beheading Holofernes,
Samson bringing down the house,
The Vandals sacking Rome . . .

The courtiers were enthralled.
'Unheard of!' they exclaimed. 'Unseen as well!'

But Parma's Duke spoke sternly:
'This circus is for common folk!
Have you no power to raise our minds
With shows of nobler deeds,
But only rapine, death and ruin?'

Faust bowed his head in shame,
And packed his raree-show.

At the gate the major-domo waited.
'His Grace desires to see in private
Adam with his bride before the Fall,
Mars and Venus caught in Vulcan's net,
King David and his virgin side by side,
And Cleopatra's nights of love .. . .
He'll make it worth your while.'

'Humble regrets,' said Faust, 'the show is over.'

The political achievements of Faust

A waste of time it would be, dear reader,
To itemize Faust's political feats,
For in effect they were no whit brighter
Than his infamously vulgar leg-pulls—
Like turning a horse to a bale of hay
Or invisibly tweaking the Pope's nose . . .
Granted, he assisted the Count of X
To repel the powerful Duke of Y
With empty suits of armour and false fire—
But neither Count nor Duke was gratified:
Honest war it wasn't, for no one died.
He furnished the Z Republic with a
Computer called Rational Government.
Its rule was just, taxes were abolished,
Crime dwindled, the economy flourished—
And the good citizens were bored to tears.
Vast Indian stretches he saved from floods,
Only to see them laid waste by a drought.
When he clad all the students in silk and
Velvet, they quickly returned to their rags . . .
Thus Faust did good, as he had wanted, and
Little good came of it. Angry he grew
And embittered, but Mephistopheles,
His faithful helper throughout, merely smiled.

Faust desires to meet Lucifer

'I would dearly meet your Master.
Perhaps he'll throw more light than you can
On life's dark matters.'

'Lucifer brings light, but darkness too.
If you suppose that I am—as you say—designing,
You ought to meet my Master!
Lord of ambiguity, Shade of many a meaning,
Spectre of the spectrum, the Giver and the Taker . . .'

He found it hard himself to comprehend his Master.

'I'm the practical party. Straight and homely,
        though not insignificant.
But I expect you'll meet him one day, dearly.'

Faust finds it is later than he thought

'Not twenty-four,' said Mephistopheles,
'But twelve years only if you please.'

'The pact,' Faust shouted, 'calls for twenty-four,
As I can prove in any court of law!'

'Your calculations must have gone astray—
You used my services by night as well as day.'

'You're nothing but the cheapest sort of cheat!'
Faust cried.

'A cheat I well may be, but I'm not cheap,'
His friend replied.

'By night I slept,' Faust said, 'and so
Did you, you dog, for all I know!'

'Indeed you slept—you slept with Gretchen,
With Helen, with the go-go girl from town—

You always rose to the occasion,
Because I never let you down . . .

And if you only dreamt at night,
Who made your dreams but I?—

Oh, I worked double time all right!'

Faust asks himself what he has gained

'The fool, they say, persists in his folly.
So does the proud man in his way—
It may be folly, but it is his own.
While still man strives, still must he stray.

And if the proud man throws good money
After bad, at least this will proclaim
His disrespect for riches,
Or prove the justness of his aim.

To save his intellectual honour
The clever man cuts off his nose—
Who knows, his noseless face may grow in stature,
Or else some grander nose may grow?
While still man strays, still shall he stride.

I told this hound of hell,
This pitman paid to help me strive,
That I would learn
        why it is the guiltless suffer,
        and why the wicked thrive—

His answer was: "As you discern,
God moves in a mysterious manner—
Unlike me. Me you can understand."

But now I think that I have understood
Whether or not the guiltless always suffer,
Not all the wicked always prosper.

And I shall write this in my blood.'

Mephistopheles assures his companion
of immortality

'Why did I not give up the ghost
When I came from my mother's belly?'
Thus lamented Faust.

                    *

'But you're lucky, Doctor,
You are privileged,
You'll be remembered—

Not exactly in flowing cups,
But in story-books,
A play or two, an opera perhaps . . .'

(He saw the stage machinery running
Amok, dragons and heaven collapsing,
Killing two or three of the actors—
Yes, that was at Lincoln's Inn.

And in a town called Exeter,
One devil too many on stage
Who didn't seem to be a player—
He would enjoy that.)

'In fact a noble Engländer
The Duke of Ellington
Will put you in an operetta
With a cute mulatto Helen!

Pity about your name though,
Doesn't have much of a ring to it.
Mephistopheles now—
There's music for you . . .

But you're fortunate, Faust,
You should count your blessings—
Most lost souls don't rate an epitaph.'

Lucifer advises Mephistopheles of the device
known as sprats and mackerels

'Someone in Weimar has tacked together
An artful defence of your friend, I gather.'

It was Lucifer speaking.
He was speaking out of a cloud.

'It seems that Faustus was only striving—
And to his Lordship striving is pleasing.'

'I strove—and he succeeded!
That tale won't cut much ice in hell!'

'I think we yield to the cunning Geheimrat,
And suffer your fortunate friend to depart.'

'You want us to look foolish, Lucifer?
How can you do this to your servant?—
It's devilish!'

The cloud that Lucifer spoke out of
Grew darker. So did his voice.

'We all have our cross to carry,
But crosses are ladders.
We let the Doctor go scot-free,
To encourage the others . . .

Experto credite!'
The storm-cloud drifted away.

Faust and Mephistopheles make their adieus

'We too can move in a mysterious way,'
He ground his teeth. 'Once in a while—
So don't pin your faith on it!

I shall tear this contract into pieces.'
He did so, very slowly.
'You're free. If free's the word . . .

I scarce expect a testimonial,
But can't you make a special effort
And just for once say Thank you?'

Faust thought of how his hand had hurt.
'Thank you,' he said at last.
'I'm not entirely taken by surprise.'

'Indeed?' The other strove to look
Entirely unsurprised. 'Mind you,
You only gain what we are glad to lose—

Time, and time hangs heavy.
You'll miss me, Faust. I was your muse.
Who will amuse you now?'

The one considered things to do in time.
The other thought of things that time could do.
The stars stood still, the clock marked time.

'That's true, my friend,' Faust sighed.
'You've ruined me for other company.
I'll pass my days in some far wilderness.'

Was there no wilderness at hand?
'I'll miss you too,' said Mephistopheles.
'But then, I'm used to missing . . .'

Midstream

Half-way across the racing river
The big man groans: 'So heavy though so small—
You bring my life in danger.'

'Here is no wonder,' says the child
Behind his shoulder. 'See, you bear the world
And all its sins as well.'

The other halts. 'What credit do I gain
By this? How long my fame?
Wiser to drop you where I stand.'

'Go on you must,' the child replies.
One thing at least he knows for certain.
'I was not born to drown.'

The other sighs: the world shall have its ride
Then, here's no place to bandy words.
He bends his strength against the flood.

K. on the Moon

'If his parents love him so much, why don't they give him 30,000
gulden so that he could go off to some cheap little place on the
Riviera . . .'

—Max Brod to Felice Bauer

Why is K. going to the Moon?
Because he is the thinnest person he knows
Because he neither smokes nor drinks
Because he does not care for communal life.

Off lifts the tin-beetle
Trailing its fiery slime
It looks good
(Except you can't see it).

But K. can see Fräulein Bauer
He can see Kakania, Krim and Korsika
Kreta, Kapkolonie and Kanada
They all look much as usual, only quieter.

'Dear Fräulein Bauer
I am an erratic letter writer—'

The typewriter floats above his head
Such things never happen in the office.

'Yet a letter is better
(Especially fine handwriting today, not so?)
If I arrived in person you would find me
Insufferable, and the thinnest person you knew.'

His burden he knows is weightless
But does his burden know?

'Will you marry me, dear Fräulein?
Will you marry a man who is sickly, taciturn, miserable
  and selfish
A man with white hair, poor prospects and palpitations
A man who is chained to invisible authorship
A man for whom even night is not night enough
A man who lives on the Moon?'

145

And would he want to marry a dear Fräulein—
Or anyone else who was willing to wed him?
Engagé he is but his health
Is not good enough for marriage.

'For a whole week I have seen nothing but
Outlandishly dressed Montenegrins in my sleep
Which gave me headaches . . .
When I recover my typewriter
I shall write a story called The Man Who Disappeared.'

He does not have a bent for writing
If you please, he is bent with writing.

'You read my letters carelessly. I said my uncle
Was in Madrid: you have moved him to Milan.
Is it all one to you whether I am on the Moon
Or on Mars . . .?'

The roads are long and it is easy to go astray.
He has acquired great experience in complaining.

'What have I done that you torment me so?
No letter today, no letter yesterday
And I believe it is now tomorrow.
Some diabolical official is playing about
With our letters . . .'

He runs in circles like a squirrel in a cage.
What might he not accomplish if he were free
And in good hands?
(Nothing.)

'Love me, Felice! And hate me!
Why do I torment you so?
You are my own self . . .
You are my human tribunal.
Please send a rose in forgiveness.'

*

A small step for mankind
But a giant leap for K.—
He falls on his face on the moon-rocks.
Have no fear, Felice—
Devised solely for writing, he cannot break.

Alas, the Montenegrins are there already
In their headachy garb!
A long thin butcher's knife glints in the moonlight.
Someone has been telling lies about him.
He is virtually a vegetarian.

We have a problem
(We always have a problem).

'Dearest F.
Now quickly forget the ghost that I was.
Peace will require no letters.
Try to drink less tea in future . . .
Kisses in the corner—K.'

The Moon is made of literature.

I am a Bewohner of an Elfenbeinturm

It seems I live in an ivory tower
Or if I am German (am I a German?)
An Elfenbeinturm is my abode.
Do I abide in an elephant's leg?
Am I a flea then, a Faustian *Floh*?

If I were Chinese (but then I am
Not, China is not in Europe as yet)
I might reside in a Jade Pavilion.
Were I a more dynamic dreamer
I might inhabit a marble hall.

But if I am German only in part
(My German is only partly good)
Perhaps I dwell in an elf's left leg.
Or Number 11 Leg Towers maybe—
It has the ring of authenticity.

That I live at some remove from life,
Its down-to-earthiness, its low *Geschmack*,
Is owing to lifts forever out of service,
Trash on the stairs, the price of real ale,
And (oh!) the cost of public transport.

But write to me (if you can spare the postage)
At Flea Flat, Leg Towers, Elephant and Castle,
And tell me about the real and earnest.

Poetical Justice

It will be many years before I read again
Of the death of Cordelia,
Or indeed (though he deserves cuffing)
Of the Macduff boy's stabbing.

(Instead I shall read of the dissolution
Of science-fictional monsters,
Or the gonorrhoea of modern heroines.)

That such things happen in life is no cause
For them to happen in literature.
If it is true that 'all reasonable beings
Naturally love justice',
Then where shall they hope to find it?

I prefer to hear of such unlikely events
As Hermione surviving in private, or
Isabella furnished with a ducal husband.
(Some have tried, too late, to save Cordelia.)

In life we barely choose our words even,
Only those we hurt will still recall them.
Art, they say, continues life by other means—
How other are they?

From *Instant Chronicles: A Life*

Entertaining women

In a night-club in Hiroshima,
A combo playing noisily,
A girl asked sweetly, *Kohi shimaska?*:
Should they make coffee?
No, he replied, it kept him awake.
It was *koi*, it struck him later, not *kohi*:
It was love she had offered, not coffee.
The thought kept him awake.

Next day, as a guest of Rotary,
He conveyed (without authority)
Fraternal greetings from Cradley Heath.
Waiting outside was a victim
(Rotary does not entertain women),
A victim for him to see, to see him.
Him with his face still scarlet,
Her with her white scarred arms.

Night Life

The busy bells of the trishaws
Chirp in the velvety night.
Here is a house of little fame,
Blue movies in black and white.

A woman attends the projector,
Ancient and prone to fail.
A baby is slung about her . . .
It's hard to make head or tail—

Surely the film is upside-down?
But no one dares to grouse.
Her husband is a policeman,
This is a private house.

All of a sudden the sound comes on,
The first sign of zest—
The baby is taking its pleasure
Sucking its mother's breast.

Armies

As if so many princes weren't enough
They fashioned generals, often out of
Sergeants from colonial legions.

'Good iron isn't used for making nails'—
The people's wisdom sometimes fails.
Nobles are born, soldiers are made.

(He'd known a minor noble working in a
Filling station: 'Blame it on polygamy.'
Also a princess who taught philology.)

The women all looked like princesses.
Virgin corporals in battle-dresses,
They learnt the ways of the soldiery.

Dense with relationship, these sudden armies
(Be careful not to rape your sister!).
In extenso died the scattered families.

Still many princes left and many generals,
Removed to milder spots where rank still tells.

Off limits

Ponto-chō, heartland of the geisha, in Kyoto: a narrow street of
wooden houses. One evening a jeep drew up at the open
entrance to a little bar. Come out, ordered the US military police:
Let's see you. To which our hero stoutly replied: You come in
here. No, said the MP: It's off limits. (Not that the US Army had
declared the bar off limits, but the bar had declared itself off
limits to the US Army: there was this reciprocal arrangement.)
Come out here, the MP repeated. Said he, insouciant: You can't
arrest me, I'm English. A muttering under the breath, and away
the jeep bounded, along the bank of the gently flowing Kamo.

But those dusty charming towns of long-gone Indo-China. Shady trees and hotels, peaceable slow-moving people, given to neatness (and to their neat children) and not-too-strenuous pleasures, graceful in their mild disgraces, a civilizing influence. Trishaw drivers chatting in French; now learning English, language of the future, a millennium of opulent fares. You can't arrest me, I speak English.

Process—how cruel it is. Nature abhors the vacuums we create, fills them with more like us. Filled women, cradles filled, streets filling up again. Ponto-chō stands where it did, a nest of singing girls. But gone are old Phnom Penh, Saigon. Come out; off limits. Such vanishings, such bans . . . like age—a kind of house arrest.

Wisps

But these little ones are so very ignorant,
It is hard to know where to begin with them.
Mostly there is something to build on, to put to rights,
Some special and meet desire—how we run to fulfil it!—
Something to complete. Here there is everything to complete.
All they wish for, impossibly, these wisps, is to be no more
    than they were, no other.

They block the roads. What realm are they subject to?—
So skimped, unlicked, such blank unwritten pages.
Folklore tells of a novice in hell, facing examinations,
Who burns exemplary answers and swallows the ashes.
They have swallowed ashes already.

What's to be done? While we ponder (we have to ponder?),
Let them be placed in—to borrow the word—in a camp.
For their own good, surely. (It has to be better than good.)
In a happy land, it goes without saying! Where sun shines
    softly, rain is gentle, peace drops slowly.
It isn't easy, best will in the world, in the best of worlds,
To provide for such guests, for these tiny ghosts.

Spring clean

'If my devils were driven out, my angels too would receive a mild shock.' One takes the German's point, at once both poetic and prudent. But might perhaps the angels, mildly shocked, be less stand-offish then? If one cried, at most they murmured 'There, there', always having something better to do. They were rarely at home; one wasn't really at home with them. Immaterial seemed the word. Once or twice one may have entertained them unawares.

And afterwards, would the devils slink away, tails between their legs? Or merely strut more insolently? He could hear them jeering: 'Give a dog a good name, it'll still bite.' Devils were probably irredeemable.

Yet, oh to fall in love with a beautiful psychiatrist! With grey eyes, long neck, chaste white smock, motherly bosom. A wolf-man he! A ministering angel she! To take her away from it all. Or would she soon grow bored? (The *Standard Edition* of SF reads like a novel—like sf in fact.) Even though he told all. And it didn't do to tell women too much.

Or men, which most of them were. Flat-chested father figures. Or school doctors: 'More outdoor sport for you, young fellow!' Devils v. Angels at basket-ball? How politely Rilke declined the offer of 'a great clearing-up'. Heine would have emptied a chamber-pot over their dingy beards.

Better the devils you know . . . till the last dim beating of that other angel's wings.

Posterity

Posterity was always a great reader.
He would beg, borrow or steal books,
He would even buy them.
You could be sure to find Posterity
With his nose in a book.
(Except when listening to music
Or peering at paintings.)
He had excellent judgement too.
You could always put your faith in old Posterity.
We shall miss him.

Meeting writers

In unaccustomed comfort
Writers were travelling abroad to meet writers.

They met with experts on writers—
On the Writer in his Time
Or in his Place
On the Writer in Society
Or in Isolation . . .

> The writers did not meet with writers—
> Perhaps they were not nice people to meet
> Perhaps they did not care to meet writers
> Perhaps the writers had unwittingly passed them
>   travelling in comfort in the opposite direction?
> Or perhaps they were actually writing.

In any case writers convey an aberrant impression of writers.
Experts are always more trustworthy.

Fathers

Visiting Belfast, he would hardly have boasted
Of how his father boasted of a Fenian father.
Remembering rather that many years before
His father was taxed with the small explosives
Stuffed into pillar-boxes—
The boxes that it was his father's job to empty.

Small cultural objects

'America, you have it better,' Goethe observed
Plaintively. Certainly better than old Poland.
For instance, the carefully faded elegance
Of the Hotel Europejski, once the pride of Europe,
Its marble stairs redeemed by a ragged carpet.
In the foyer, behind large slices of plate glass,
Small cultural objects cower, for sale or not,
And a native poet confides to a foreign one
That three poems have been banned from his new book—
Should he withdraw it? Or is half a loaf better . . .?
An enormous radio is transfixed to one transmitter:
All the better to be heard. Outside the chamber
Sits a deadpan chambermaid. Doubtfully for sale.
Servants are not servants, they are civil servants,
They wish to meet your wishes, which had best be modest.
Back in the far-flung shadows of the restaurant
A cluster of rustic hussies try to remember
What cocottes were like. The Palm Court players
Explode in a waggish folksong, and they stop trying.
Waiters whisper hungrily, something about dollars.
Are they provocative agents, like the hussies?
Here are the grand menus still (stained with bortsch),
But not the dishes. Half a loaf . . .

Where can it be?

The seat of the soul. Hardly in the purlieus given over to
pleasure. Possibilities were struck off one by one: the nose,
blocked with grime, prone to offend its owner; the eye, ever
looking for offence; the all-too-busy hand. (How can it be, as a
modern sage declares, that the body is the best picture of the
soul?) Until only the zones of pain remained. Should one suffer
solely from gout, would the soul therefore reside in the big toe?
Well then, the breast, or the heart? Once highly desirable
neighbourhoods but now overrun by lawless pangs and migrant
uneases. Or the head, the capital itself? Smoggy, littered with
grubby worries, addicted to aspirin and other drugs. Or, as
some holy man surmised, the pineal gland? A 'pea-sized organ
in the brain'. Roominess was immaterial, but even so, not a very
grand address. If no suitable seat could be located—would this
suggest there was no soul?

The evil days

While the years draw nigh when the clattering typewriter is a burden, likewise a parcel of books from the postman, and he shall say, I have no pleasure in them; for much study is a weariness of the flesh;

Also when the cistern shall break, and the overflow be loosed like a fountain; when the lights are darkened, and the windows need cleaning,

And the keeper of the house shall tremble at rate bills, and be afraid of prices which are high; and almonds are too much for the grinders, and beers shall be out of the way;

Yet desire may not utterly fail, and he shall rise up at the sight of a bird, when the singsong girls are brought low;

In the day when he seeks out acceptable words; when the editors are broken at their desks, and the sound of the publishers shall cease because they are few.

Then shall the dust return to the earth as it was, and the spirit also, whether it be good or evil, shall look for its place.

Last things

You need to be a Catholic to stand that sacrament
Going on around you. Otherwise, passionless faces
Of nurses, and someone you vaguely know, saying
The worst is over and you'll soon feel much better.
Probably you will. It can't hurt to leave with the
Sound of truth in your ears. Unless you guess at
Rewards or reprisals ahead, and doubt that better's
The word for you. In which case the last rites are
Not so bad. (Extreme unction has helped you before.)
Nor, come to that, is the needle, that blunt visa.
At least go in peace, since you must go, in peace.

Peace, here, being primarily a physical affair. This is the body's hour. The soul, wherever situated at the moment, has all eternity before it. It's the flesh that wants to be put decently to sleep. The spirit, with a care for residual dignity, will not disapprove. May even concur. They were closer, over the years, than was often supposed.

Ever-rolling

Though much is uttered at many assemblies
We fail to hear the holy voice of Clio.

History perhaps has ceased to make herself;
More than enough, she thinks, to last for ever.

Like brilliant fog, the dead, the famous ones,
Eat up our oxygen, and nod politely—

At every man a king, king for a day, then
Lost in a common last unfinished phrase . . .

Those quivering mounds of flesh and tatters—
Matter for some new Gibbon or Macaulay?

A score or more of leading international poets
Mount a platform. The earth declines to move.

Some thought they saw the shade of Thomas Mann
Linger by a news-stand, a pained look on its face:

'After me a deluge. Merely a flood.'
Perturbed of course, and yet a trifle smug.

While ever-rolling streets bear shades and
Solids, agents, patients, all away,

And still we sit in chains, where film
Unwinds its strings of undistinguished horrors.

If all the innocents, the slaughtered commons,
Swollen children, if only they could rise

In one great cloud to heaven, at least
The guiltless, could go and leave us free,

Leave us a yet unwritten page. A virgin year,
And this tired soil lie fallow. What might follow?

We thought we prized the past, its noble gifts,
White elephants that ate us out of heart and home.

If on the moon mankind could lose its memories;
On some fiery star our brains be wiped quite clean.

Until which time we make our unfresh starts
And share our instant chronicles. It's your turn now.

An Old Story

There were to be a thousand of them, Buddhas or something
  akin,
And every one of them different. He had finished 999,
And they were all and each, in some particular, different.
He had worked as fast as he could, but time was running out.

Now they came puffing up the hillside, abbots and priests,
Provincial governors and the usual lackeys; also (he feared)
An executioner from the Board of Punishments, severe of
  mien.
He strove to imagine the thousandth and its peculiar
  uniqueness.
They were just round the corner, the tally-man was shouting
'990 . . . 995 . . . 998 . . .' The executioner pursed his lips,
While the priests and the governors shaped the beginnings of a
Discreet smile. The Emperor was a stickler for punctuality.

The niche at least was ready. He jumped into it, cocking his
  head
And raising his hands in what he hoped was a brand-new
  position—
With luck they wouldn't notice that the thousandth was
  alive . . .
It wasn't. Some benign and witty divinity turned him into
  stone.
The sculptor was sculpture, the master his own masterpiece.

Much later came the new, the young, the red-blooded,
  licensed
To renounce the past and all its works. Ancient philosophers
Had been tried and condemned *in absentia*. Recent poets
Were sent to clean out latrines. The watchword was 'Realism!'
And now came the turn of the thousand Buddhas or
  whatever—
Equally unreal and irrelevant—they were. Crash, crash went
The sledge-hammers, demolishing the minute particulars and
Reducing uniquenesses. There was much to be done and
  undone.
They hastened to the thousandth. Crash went the hammers.

See, he has lost his head. But otherwise he is all there.
Which can hardly be said of some among us standing here,
Encouraged to gawk, now that policies have been slightly
    reversed
And the past is permitted. And we mutter ruefully an apt
And innocuous proverb: 'Easier to pull down than build up',
Cocking our heads, unsure what to do with our hands.

Mixed Proverbs

Outside the restaurant, garish but inside delicious no doubt,
A most famous of Guangzhou, if not of Canton as once,
Excited citizens are stringing up garlands of firecrackers.
What they are arguing about is beyond the visitor's
    knowing—
When he offers his lighter they absently wave it aside.
It may be a question of how best to arrange them, or whether
Such acts are utterly licit. *The people fear the officials*,
Thinks the foreign devil, who has heard of this local saying.
For how to tell what the party line currently is on fireworks,
Their potency and periodicity, their season and station?
He remembers his childhood, and a miserable sprinkling of
    them,
They didn't have money to burn, and the sense of dull guilt
When a rocket winked out or a Catherine wheel stopped
    dead.
Come on, he wants to cry, let's have them, a good half-hour
Of continuous din, driving out evil spirits (should any
    remain),
And damn the expense! But the bickering goes on, matches
    are
Struck, then pinched out, fingers point upwards and
    downwards.

His official hosts, at the door, solicit him nervously, please to
Enter, the feast is afoot. *The officials fear the foreign devils.*
What the eye never saw, the heart never grieved over; he
    can't
Pretend he ever yearned for stewed turtle or sea cucumber.
But he thinks of the engine his father once bought for him,
And even a tender. And somehow a section of rail was
    wanting,
The circle incomplete: when they forced the pieces together
The engine keeled over. The open floor was the only place
    for it,
Unauthentic and pitiful, as it banged into chair-legs and feet.
No matter, he's off to Shanghai tomorrow. It's a big country,
It can't be short of railway track. Sometimes the years supply

Our childhood lacks. All things come to them that wait.
But he must make sure to seem small and humble, and
  kowtow
To the woman who keeps the key to the WC.
The foreign devils fear the people. Now the circle's perfect.

Writer's Tinnitus

So it's come to this! Not only when you've
Just brought out a book, and not quite
Faintly praising: more like a hissing,
Not especially faint, and minus respite.

In sweeter terms, it's what you still remember
Fondly: the long continuous chirring
Of cicadas, through the night, in one or other
Long-left place—that tropic velvet,
That fecund heat—making a second, more stirring
Sort of silence. But what enhances darkness
By day grows merely stupid.

Or you might say, a distant soughing,
A sea whose waves unbrokenly are breaking;
Or some well-mannered but unfaltering wind
Through brittle grass; or gas escaping.

Or tittle-tattle's buzz: a new biography
In the offing, rending of veils and shrouds?
(For truth is great and shall prevail,
If stiffened with pornography.)

Or else, at speed, a tape of all the words
That ever were, stripped of all their affect.
Tools of the trade reduced to rusty susurration,
Ancient arid bookstacks calcining unchecked.

But this is mere imagination
(A presence which itself is hissing, darkly
Whispering in the wings, without remission,
And lucidly, alas)—
And you no better than the class of ass
Who bores his doctor with prodigious symptoms!

Still, something's cooking. A pullulation
Of micro-organisms, non-readers all the lot,
All ranters? Then it's when the ululation
Ceases that you start to worry. Or you stop.

Failing

Almost stone deaf she is now, unless
Preferring to hear nothing further
Of our sorry world, all victims or crooks.
Tired, I guess, of people's cleverness,
As mine must once have fretted her.
She has had her share of cares.
She listens, she tells me, to Talking Books.
I look at her tapes: *The Sea*, *The Sea*—
Oh mother, I was never clever like that!
I recall her saying: 'That Pam Ayres,
Isn't she doing well with her poetry!'
(Something for others to wonder at.)
Now her eyes are failing, can't make out TV.
She's had enough, of woes, joys, scares,
And for today, at least, of me.

From *Monkey Puzzles*

Monkey recurs

He has been bad again.
Vanity, anger, drunkenness, tattle,
Avarice—he coveted the old age pension—
And concupiscence. In a nutshell.
Also a brief bout of sloth
Wherein he committed very little.
Bemused by a high-slit skirt
He travelled three stations without a ticket;
Or was it a low cleavage?
His memory is bad as well.
He leafed through dirty books in a shop,
And didn't buy them: meanness is wicked.
He abstained from taking life,
And instead took umbrage.
Seeking to purge and to purify
He came unstuck through attachments.
Buddha is not pleased with him.
'Egotism, envy, spleen and lust,
The iron itself creates the rust.'
Dharma may not be denied.
'Bound to the Wheel' is the Lord's sentence:
'The whole thing all over again!'
Thank you, thinks Monkey, hanging his head.
(He was good at concealing pride.)
He remembers to groan a piteous amen
As befits the justly convicted.
Nor would he wish to be thought a Nietzschean—
At least he had never envied Superman.

On his rights

Monkey rights are monkey business,
Simply simian. We cook our geese
And settle our hash.
We are Primates inter pares,
Our grunts are full of grace.
We don't ape others. Don't ape us.
It's nuts to you.

165

OXFORD POETS

Fleur Adcock

James Berry

Edward Kamau Brathwaite

Joseph Brodsky

Michael Donaghy

D. J. Enright

Roy Fisher

David Gascoyne

David Harsent

Anthony Hecht

Zbigniew Herbert

Thomas Kinsella

Brad Leithauser

Derek Mahon

Medbh McGuckian

James Merrill

John Montague

Peter Porter

Craig Raine

Christopher Reid

Stephen Romer

Carole Satyamurti

Peter Scupham

Penelope Shuttle

Louis Simpson

Anne Stevenson

George Szirtes

Grete Tartler

Anthony Thwaite

Charles Tomlinson

Chris Wallace-Crabbe

Hugo Williams

also

Basil Bunting

W. H. Davies

Keith Douglas

Ivor Gurney

Edward Thomas